ASHEVILLE-BUNCOMBE TECHNICAL INSTITUTE

FUNCTIONAL GAGING

DISCARDED

JUN 24 2025

FUNCTIONAL GAGING

Second Edition, Revised

EDWARD S. ROTH
Member of the Technical Staff
Sandia Laboratories

SOCIETY OF MANUFACTURING ENGINEERS
Dearborn, Michigan
1970

Functional Gaging

COPYRIGHT © 1970 by the
Society of Manufacturing Engineers,
Dearborn, Michigan

COPYRIGHT © 1964 by the
American Society of Tool and Manufacturing Engineers,
Detroit, Michigan

All rights reserved, including those of translation. This book, or parts thereof, may not be reproduced in any form without permission of the copyright owner. The Society does not, by publication of data in this book, ensure to anyone the use of such data against liability of any kind, including infringement of any patent. Publication of any data in this book does not constitute a recommendation of any patent or proprietary right that may be involved.

Library of Congress Catalog Card Number: 74-118771
International Standard Book Number: 0-87263-019-6

SECOND EDITION, REVISED • MANUFACTURING DATA SERIES
MANUFACTURED IN THE UNITED STATES OF AMERICA

PREFACE

The second edition of this book adheres to the original purpose of using actual gage illustrations to explain each functional gaging concept.

Many parts of the book have been completely rewritten, symbols have been revised, and the appendix material now includes selected portions of American National Standard Y14.5, "Dimensioning and Tolerancing for Engineering Drawings," which replaces MIL-STD-8.

Acceptance of the positional tolerancing concept has increased markedly since the first edition of this book was published in 1964. It is now estimated that well over half of the aerospace industries now use the concept. Design clarification along with attendant cost savings have been the obvious benefits.

The publication of ANS Y14.5 in 1966 has been of great value in promulgating acceptance of positional tolerancing because this standard reflects the comment and review of American industry and military establishments, and has been coordinated with British, Canadian and The International Organization for Standardization. It has been most satisfying to note the progress made in documentation and use of the true position dimensioning and tolerancing concept these past years. The concept is timely, and its complete acceptance is only a question of time.

The author is indebted to the officers and staff of the Society of Manufacturing Engineers for their foresight in publishing the first edition and in this edition to update the concept. It has been of great benefit to participate in the documentation of this original material.

EDWARD S. ROTH

PREFACE TO FIRST EDITION

The publication of several international positional tolerancing standards during the past decade has directly affected product, tool, and gage engineering. Military Standard 8 and British Standard 308 were published in 1953 and the Canadian Standard was published several years later. The subsequent use of true-position dimensioning from well defined "datums" (planes or axes) and positional tolerancing specifications have forced the designer to think ever more carefully and to define many geometrical relationships that he had not previously considered. Design ambiguities, however, still plague tool and gage engineers when parts are dimensioned from implied "datums" with bilateral *or* positional tolerances, and the detrimental effect on product acceptance rates can probably never be determined. The positional tolerance system can give the designer absolute control over final product quality by dictating the specific gage design criteria that determine functionally acceptable parts.

This reference text should help manufacturing engineers determine tool and in-process gaging requirements; it should help gage designers determine the form, relationship, and location of gage elements; and it should help product designers dimension and tolerance engineering drawings so that practical tools and gages may be used.

The true-position dimensioning and positional tolerancing system has enabled designers to substantially increase tolerances. Datum-dimensioned bilateral tolerances can be increased over 50 percent by conversion to positional tolerances. The use of the Maximum Material Condition (MMC) modifier will usually increase positional tolerances approximately 100 percent over bilateral tolerances when part features are finished to nominal size.

Further cost savings can be effected if designers use 0.000-in. diam positional tolerances at Maximum Material Condition. This technique does away with separate Go plug or ring gages because these are included in the receiver gage when this callout is used.

This text limits itself to "functional gages"; that is, gages that "receive" the part being inspected, much like a mating part, and which contain fixed elements (pins, bushings, etc.) to check part features. Functional gages can easily be designed by direct mathematical methods if parts are defined from specified part datum-features with true-position dimensions and positional tolerances. Arbitrary non-mathematical methods must be used to design similar gages, however, when the bilateral tolerancing system (i.e. square or rectangular tolerance zones) is used to define the product.

This text uses the symbols and specifications in MIL-STD-8 (*see* Appendix B), but goes beyond the scope of this Standard in the application of "datum" dimensioning and variable positional tolerances requiring fixed-element functional gages. Some nonstandard dimensioning methods, notes, and symbols have been added as necessary so that practically designed gages can be used. It is not suggested that these nonstandard methods should become standard. They are added for the convenience of the reader, and indicate specific points that must be covered if drawings are to completely define the design intent. The author uses MIL-STD-8 symbols because they are close to an "international language."

Specific and simple illustrations have been purposely developed to directly show single gaging principles. The basic principles of true-position dimensioning and positional tolerancing theory are only briefly mentioned since they are adequately covered in related standards such as MIL-STD-8 and Process Standard 9,900,011 (*see* Appendix A). Only those methods of dimensioning and tolerancing are considered that lead to simple and practically designed tools and gages. Complex moveable-element gages containing sliding or diamond pins are used in industry, but indicate questionable dimensioning, tolerancing, and tool design techniques since gages should usually simulate mating parts.

Apparently due to incompatibility with bilateral tolerancing, no discussion of fixed-element functional gages has been published. Instead, these "illegal" principles have been kept in personal "black books" or company standards. Fortunately, these gaging principles are sound and are "legal" when positional tolerances, modified with Maximum Material Condition, are specified on product drawings. Gage tolerances and wear allowances are not covered in this text since these practices are detailed in MIL-STD-120.

True-position dimensioning and positional tolerancing may well become the international system of product definition. This will come about when the standard on this subject is published by the International Organization for Standards (ISO). Most of the manufacturing countries (Great Britain,

Canada, France, Germany, Switzerland, Italy, the U.S.A. and the Communist Bloc) belong to ISO and contribute to the Standard. The ISO Standard has not been included in the Appendix because it is still in preparation. ISO Symbols are, however, included in Figure 1-3 in the text.

Both positional tolerancing and fixed-element functional gages were in use long before the author's twelve years of experience began in this field. Numerous consulting and teaching assignments have, however, enabled him to examine, compare, and then compile the drawing-gage relationships and principles presented. Thanks are due the management of Sandia Corp. for encouragement and support, and many former students for the use of their ideas, advice, and assistance.

I dedicate this book to my wife, Pat.

EDWARD S. ROTH

CONTENTS

PREFACE	v
1 DEFINITIONS, PRINCIPLES, AND STANDARDS	1
2 FUNDAMENTALS OF POSITIONAL TOLERANCING	9
3 DESIGN PRINCIPLES FOR FEATURE RELATION GAGING	30
4 DESIGN PRINCIPLES FOR FEATURE LOCATION AND RELATION GAGING	43
5 GAGING FORM TOLERANCES	78
6 SELECTED PORTIONS OF ANS Y14.5, DIMENSIONING AND TOLERANCING FOR ENGINEERING DRAWINGS	83
BIBLIOGRAPHY	133
INDEX	136

CHAPTER 1

DEFINITIONS, PRINCIPLES, AND STANDARDS

The gages discussed in this book function as mating parts. Known as functional (or receiver) gages, they simulate the most critical conformation of the mating part when they "receive" the part being gaged. They do not give variables inspection data, but only tell the user if the part is accepted or rejected. The text will discuss the two types of functional gages: (1) feature relation and (2) feature location and relation. Such instruments have a fixed configuration, like the mating parts they simulate, and allow each part gaged a different tolerance since no two parts can ever be identical in size and form.

Non-variable tolerancing methods, such as bilateral tolerancing or positional tolerancing when *not modified with the Maximum Material Condition* callout, are not compatible with functional gages. Only the advent of the variable modifier, Maximum Material Condition (MMC), which specifically allows part feature locational and form tolerances to vary with part feature size, enabled this mathematically sound text to be written.

The parts gaged in this book are defined by a standard. The gages are not. Although it is not accepted practice, there is no reason why the ANSI Standard which describes datum requirements, geometrical characteristics, and tolerances, should not apply to gage definition as well.

DEFINITIONS

Basic (BSC). The perfect (and desired) location of a part feature. Interchangeable with the term "true position."

Concentricity Tolerance (Coaxiality). A cylindrical tolerance zone, concentric to a datum axis, within which the axis of a feature must lie.

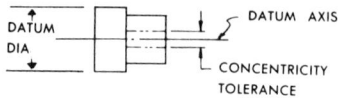

Profile Tolerance. A zone of specified width and depth along or about a basic three-dimensional part boundary within which the actual finished surface must lie.

Datum Planes. Mutually perpendicular datum planes established by gage datum elements associated with designated part reference surfaces. Reference planes from which measurements are taken.

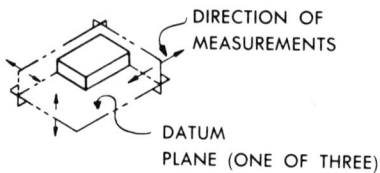

Datum Axis. The intersection of two mutually perpendicular datum planes established by gage datum elements. A reference axis from which measurements are taken.

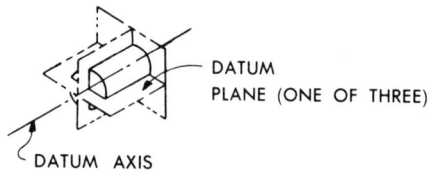

Datum Allowance. A difference in size between the gage datum element and the part datum-feature at MMC. A specified gage datum element fit tolerance.

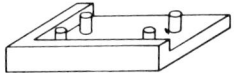

Feature Location and Relation Gage. A gage that checks the location of a part feature, or a pattern of part features, relative to a part datum-feature. Such a gage will contact two or more part datum surfaces.

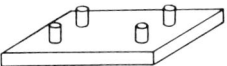

Feature Relation Gage. A gage that checks only the relationship of a pattern of part features. Such a gage will contact only the primary datum surface on the part.

Fixed-Element Functional Gage. A gage that simulates the form of the mating part; a "shake" gage; a receiver gage.

Fixed Gage-Element. Any pin, bushing, or surface on a gage that cannot be adjusted for size or location.

Fixture Gage. A functional gage that contains indicators or other similar devices.

Gage Centering Element. Any gage datum element that centers on a part datum-feature.

Gage Element. Any portion of a functional gage.

Gage Shake. The actual difference between a gage datum element and a finished part datum-feature.

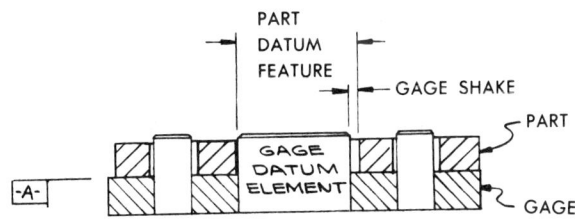

Go Gage. A plug or ring gage made essentially to the MMC size of a part feature and which is sized to reject any part feature at MMC.

Maximum Material Condition (MMC). The maximum specified limit of a size of an external feature, or the minimum specified limit of size of an internal feature. The most critical specified interchangeable size of a part feature.

Minimum Material. The minimum specified limit of size of an external feature, or the maximum specified limit of size of an internal feature. Usually the least critical size of a part feature.

Not-Go Gage. A gage made essentially to the minimum material size of a part feature and sized to reject any part feature at minimum material.

Datum Feature. A designated part "datum" surface or feature (hole, diam., etc.), usually qualified with datum targets or form tolerances, that contacts the gage datum elements. Dimensions are shown from part datum features.

Primary Datum Feature. A part datum surface usually contacted at three points by the gage datum elements (Fig. A7, Chapt. 6, Appendix *A*).

Secondary Datum Feature. A part datum surface usually contacted at two points by the gage datum elements (Fig. A7, Chapt. 6, Appendix *A*).

Tertiary Datum Feature. A part datum surface usually contacted at one point by the gage datum element (Fig. A7, Chapt. 6, Appendix *A*).

Part Feature. Any surface, hole, boss, pin, tapped hole, slot, etc., on a part.

Positional Tolerance. (a) A cylindrical zone within which the actual centerline of a finished part feature must lie (Fig. 2-10), or (b) a zone defined by two parallel planes, within which the actual center plane of a finished part feature must lie (Fig. 2-11).

Tolerance Zone Projection. The specified height of a tolerance zone that projects from a datum surface, within which the actual extended center line of a finished part feature must lie.

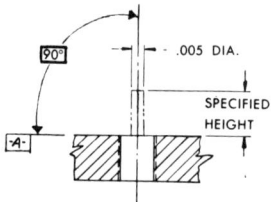

A circled capital letter "P" is the symbol used to specify the maximum thickness of the mating part (and therefore the gage bushing height) when specifying tapped hole positional and perpendicularity tolerances.

Regardless of Feature Size (RFS). The actual axis of a part datum-feature must be used regardless of its finished size (within specified limits). A gage datum centering element should be used to contact the part datum-feature. When applied to certain positional and form tolerances, the stated tolerance must be maintained irrespective of the size variation of the finished part feature.

Straightness Tolerance. The area between two parallel straight line elements, within which all elements of an actual finished surface must lie.

Chap. 1 *Definitions, Principles, and Standards* 5

Squareness Tolerance. (a) A zone defined by two parallel planes perpendicular to a datum plane or axis within which the actual surface or center plane of a finished part feature must lie, or (b) a cylindrical zone perpendicular to a datum plane or axis within which the actual center line of a finished part feature must lie.

Symmetry Tolerance. A positional tolerance. A zone defined by two parallel planes symmetrically located with respect to a datum plane or axis within which the actual center line or center plane of a finished part feature must lie (Fig. 2-11).

True Position. The perfect (and desired) location of a part feature. Interchangeable with the term basic (BSC).

FUNCTIONAL GAGE PRINCIPLES

The following principles underline the design of every functional gage, and are stated here so that the reader may refer back to them as he uses this text.

Principle 1
Functional gages have fixed gaging elements centered at part basic feature locations.

Principle 2
Part tolerances are variables since finished part features will vary in size. It is impossible to produce finished parts that are exactly alike.

Principle 3
A fixed-element functional gage should simulate the mating part at the part-gage interface.

Principle 4
A receiver gage is only functional when it physically simulates the most critical or MMC mating part.

Principle 5
The gage designer should not have to make arbitrary decisions regarding gage element size or location, since a complete product specification dictates these design and interchangeability criteria. *The drawing is not complete if such decisions are required.* The gage designer should design only that gage structure which aligns and relates these fixed gaging elements and he should tolerance the size and location of these elements in accordance with the accepted practices defined in MIL-STD-120.

Principle 6
Parts that can be practically gaged can also be practically tooled since tools and gages should be "interchangeable." Although tools usually have clamping devices and are designed to withstand cutting forces, their form

at tool-part interfaces should be similar to the gages used to accept the same product. For instance, if drill bushings are removed and replaced by pins, the tool becomes a gage.

Principle 7

All functional gage elements should "go" into or over part features simultaneously since all part assembly features (bolts, pins, etc.) must enter mating parts at the same time.

Principle 8

If several "identical" functional gages are all within their specification limits, any part accepted by any one of these gages is acceptable.

Principle 9

One "datum" (composed of several mutually perpendicular part datum features) per part will enable one gage to be used for acceptance. Any increase in the number of "datums" will increase the number of gages and inspection setups, and will therefore increase part cost.

Principle 10

In the final analysis, inspection policy should be centered in the principle of acceptance—not rejection—of all possible in-tolerance parts.

RELATED STANDARDS

The following figures compare ANS Y14.5 symbols and callouts with other American, British, Canadian, and International Standards. Figure 1-1 shows the feature control symbol for form tolerances. Fig. 1-2 shows the

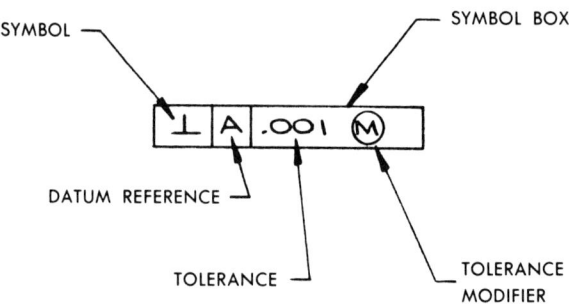

Fig. 1-1. Form tolerance feature control symbol box.

slightly modified feature control symbol used for positional tolerances. Both are found in ANS Y14.5. Figure 1-3 compares the ANS Y14.5 symbols with those proposed by the ISO. These symbols are quite similar (many are identical) and much progress has been made toward an international

Chap. 1   Definitions, Principles, and Standards

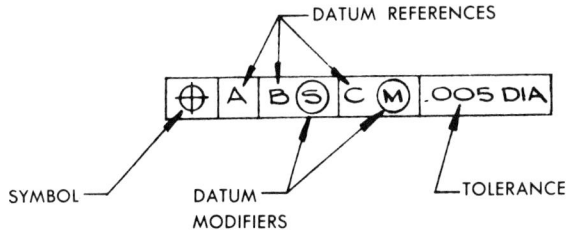

Fig. 1-2.   Positional tolerance feature control symbol box.

ITEM	ANS Y 14.5	ISO
ANGULARITY	∠	∠
CONCENTRICITY (COAXIALITY)	⊚	⊚
CYLINDRICITY	⌭	⌭
FLATNESS	▱	▱
MAXIMUM MATERIAL CONDITION	Ⓜ	Ⓜ
PERPENDICULARITY (SQUARENESS)	⊥	⊥
PARALLELISM	∥	∥
POSITIONAL TOLERANCE	⊕	⊕
PROFILE (OF A LINE)	⌒	⌒
PROFILE (OF A SURFACE)	⌓	⌓
REGARDLESS OF FEATURE SIZE	Ⓢ	—
ROUNDNESS (CIRCULARITY)	○	○
SYMMETRY	≡	≡
STRAIGHTNESS	—	—
RUN OUT	↗	↗
DATUM FEATURE	-A-	A
DIAMETER	DIA	⌀

Fig. 1-3.   Comparison of ANS Y14.5 and ISO symbols.

standard. Figure 1-4 compares ANS Y14.5 with the British, Canadian, and ISO Symbols.

The ratios between part tolerances and gage tolerances, and gage wear limits are not standardized. Therefore, gage tolerances and wear allowances must be included in contractual agreements to avoid argument when part rejections occur.

Item	International ISO/TC 10 (Symbols mandatory)	American USASI Y14.5 (Symbols optional)	British BS 308	Canadian B 78.1
			(No symbols - equivalent terms)	
Straightness	—	—	STR TOL	STRAIGHT WITHIN
Roundness (circularity)	○	○	RD TOL	ROUND TOL
Profile of any line	⌒	⌒	Not specified	Not specified
Flatness	▱	▱	FLAT TOL	FLAT WITHIN
Cylindricity	⌭	⌭	CYL TOL	Not specified
Profile of any surface	⌒	⌒	TOL ZONE	TOL ZONE
Parallelism	∥	∥	PAR TOL	PARALLEL TO
Perpendicularity (squareness)	⊥	⊥	SQ TOL	SQUARE WITH
Angularity	∠	∠	ANG TOL	ANG TOL
Runout	↗	↗	Not specified	Not specified
True position	⊕	⊕	POSN TOL	LOCATE WITHIN
Concentricity	◎	◎	CONC TOL	CONCENTRIC TO
Symmetry	≡	≡	SYM TOL	SYMMETRICAL WITHIN
Datum feature	[A]	[-A-]	⟶A	⟶A
True position dimension	[127]	[5.000]	5.000 TP	5.000 (TP)
Maximum material condition	Ⓜ	Ⓜ	MMC	MMC
Regardless of feature size	Not specified	Ⓢ	Not specified	Not specified
Tolerance	Total value specified	Total value specified except where radial (half) value is an option.	Total value specified	Total value specified
Diameter	⌀	DIA	DIA or ⌀	DIA or ⌀
Shape of tolerance zone	Zone is a width in direction of leader arrow. ⌀ specified where zone is circular or cylindrical.	DIA, TOTAL, or R specified as applicable for positional tolerances. Otherwise, not specified and diameter or width implied as applicable.	DIA or WIDE specified as applicable.	FIM or DIA specified when considered necessary.
Sequence within geometric tolerance notation	1st-Geometric characteristic symbol 2nd-Tolerance value & modifier 3rd-Datum reference & modifier	1st-Geometric characteristic symbol 2nd-Datum reference & modifier 3rd-Tolerance value & modifier	1st-Geometric characteristic term 2nd-Tolerance value & modifier 3rd-Datum reference & modifier	1st-Geometric characteristic term 2nd-Datum reference & modifier 3rd-Tolerance value & modifier

DIFFERENCES IN SYMBOLIZATION FOR POSITIONAL AND FORM TOLERANCES

Fig. 1-4. Comparison of ISO, American, British, and Canadian standards. [Extracted from ANS Standard Drafting Practices, Dimensioning and Tolerancing for Engineering Drawings (ANS Y14.5–1966) with the permission of the publisher, The American Society of Mechanical Engineers, United Engineering Center, 345 East 47th Street, New York, N.Y. 10017.]

CHAPTER 2

FUNDAMENTALS OF POSITIONAL TOLERANCING

Datum Reference Planes

A part is usually drawn in some fixed relationship to mutually perpendicular reference planes, and part features are located by means of dimensions from such planes. The reference planes may be compared with the X, Y, and Z axes of the Cartesian coordinate system (Fig. 2-1).

Part datum features suitable for contact by tools and gages (and, of course, by mating part datum features) are usually identified by letter designations. Part datum features should, of necessity, be more accurate than the part features located from them; otherwise their variations can rob the part features of a large percentage of their allotted tolerances. To be useful for tooling and gaging purposes, dimensions from "accuritized" part datum features should be directly usable without calculation.

"Datum" dimensioning (all dimensions on a part from one or several mutually perpendicular datum features) is entirely compatible with machining practices and standard tool and gage design. It is highly advisable to set up the part only once in relation to the three mutually perpendicular machine tool axes and then to have subsequent tools and gages use the identical part datum features if several operations or gaging steps are required. "Chain" dimensioning of part datum-features does not facilitate the use of one-piece functional gages since several sets of datum-features on the same part require several setups, several tools, and several gages. Unless

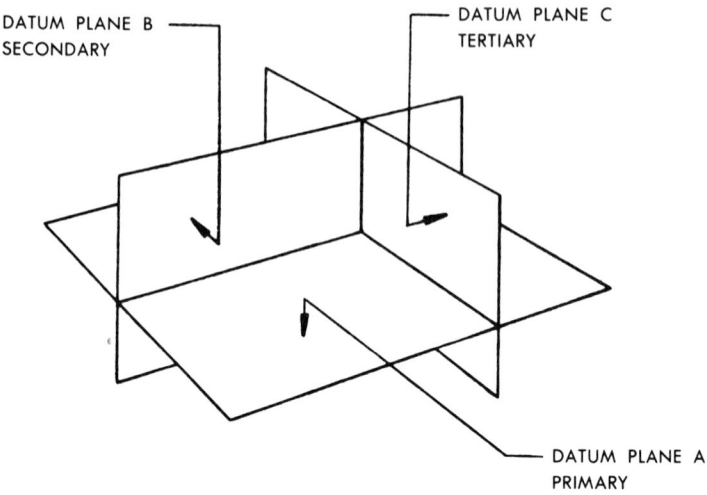

Fig. 2-1. Datum reference planes.

the part is too large to process at one setup, "chain" dimensioning of part "datums" is not practical because of higher tooling and production costs.

Part Datum Features

Figure 2-2 shows a three-hole pattern dimensioned from three mutually perpendicular part datum surfaces, each assigned a letter to show its precedence in locating the part. Part datum feature A is the primary surface that contacts the mating part, a plane established by the three highest points (Fig. 2-3); part datum feature B is of secondary importance, being established by the two highest points (Fig. 2-3); and part datum feature C, of tertiary importance, is established by the one highest point (Fig. 2-3). Theoretical datum feature reference planes are shown in Fig. 2-1, and their degree of precedence in part location and alignment is shown in Fig. 2-3 which combines Figs. 2-1 and 2-2. Assembly will force the part to contact datum reference plane A at the three highest points on part surface A. Then if the part is forced against datum reference planes B and C, surface B will contact datum reference plane B at the two highest points; and surface C will contact datum reference plane C at the one highest point.

This relationship will exist when:

1. There are three mutually perpendicular planes (three machined tool surfaces will approximate this).
2. The part surfaces are not convex, and
3. Solid three-, two-, and one-point contacts are assured through the use of tooling buttons (instead of machined surfaces on the gage) that con-

Chap. 2 Fundamentals of Positional Tolerancing

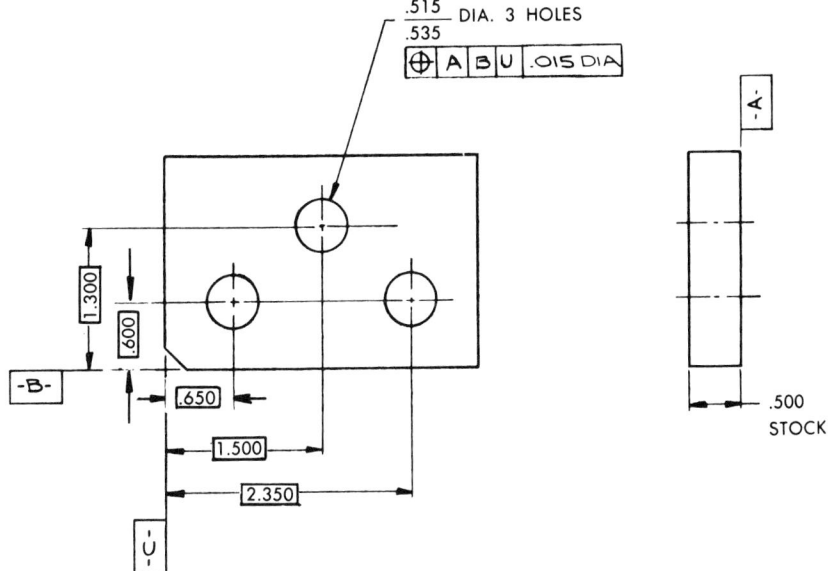

Fig. 2-2. "Datum" dimensioning.

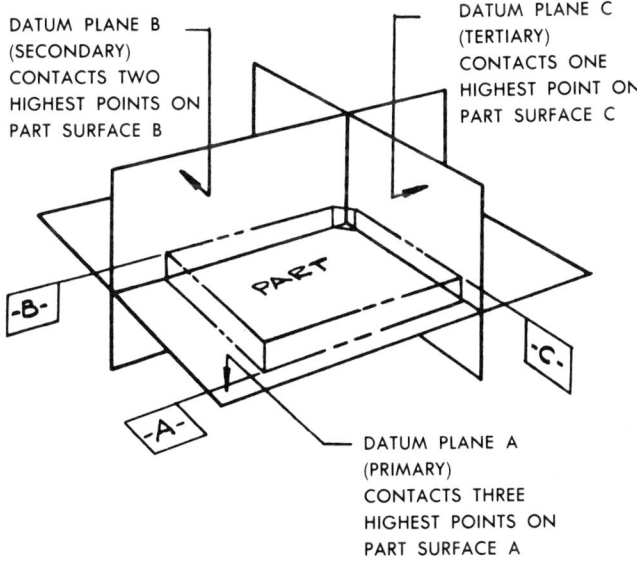

Fig. 2-3. Part related to datum reference planes.

tact the part at specified locations. Figure 2-4 shows several of the many ways whereby round and cylindrical parts can also be related to datum reference planes.

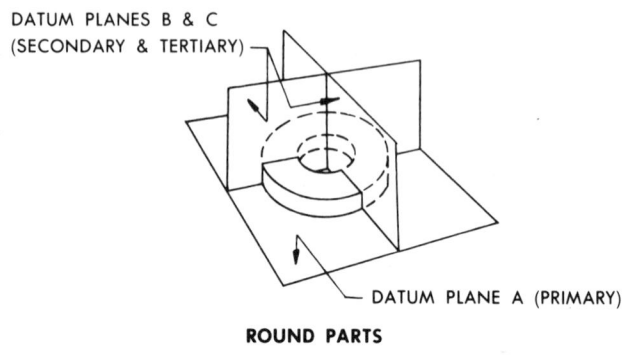

ROUND PARTS

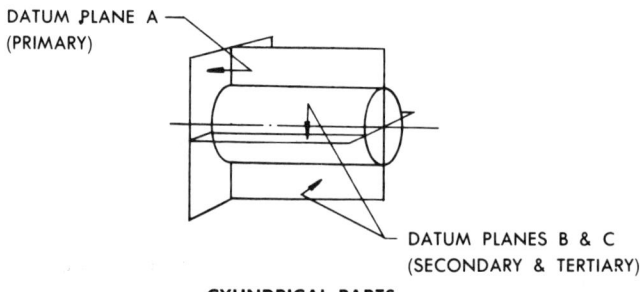

CYLINDRICAL PARTS

Fig. 2-4. Round and cylindrical parts related to datum reference planes.

Figure 2-5 shows how the part shown in Fig. 2-2 can be dimensioned from datum reference planes established by:

1. The plane through the three high points on part datum feature *A* that has 0.001-in. flatness tolerance and the stipulation that it cannot be convex (which would allow the part to rock on a machined tool or gage surface).

2. The plane through the two points on part datum feature *B*, that locates two points, with the tool and gage pickup point symbols (called "datum targets") illustrated.

3. The plane through the one point shown by the tool and gage pickup point shown against part datum feature *C*.

Gage Datum Elements

Actual datum planes, which are the origin of all measurements, are determined by the geometry of the tool or gage datum elements; that is, the actual "datum" shown in Fig. 2-5 is established by the tool or gage shown

Chap. 2 Fundamentals of Positional Tolerancing

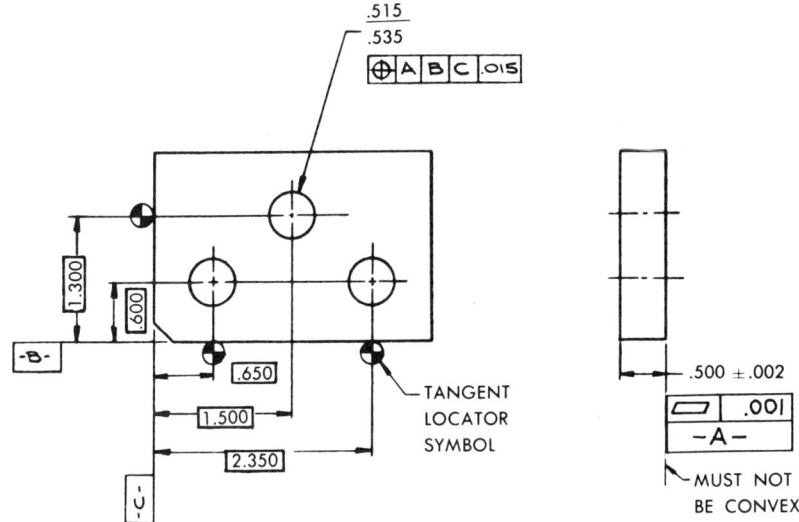

Fig. 2-5. Tangent locator symbols.

in Fig. 2-6. Part datum feature *A* is contacted by a machined surface on the tool or gage, and part datum features *B* and *C* by three dowel pins that further locate and align the part.

Figure 2-7*A* shows part datum features contacted by six tool and gage pickup points that are useful, particularly on cast and forged parts. These tool and gage points are usually placed on drawings with untoleranced dimensions, but should be correspondingly located on tools and gages to ± 0.005-in. maximum tolerances to ensure datum repeatability between tools and gages. Figure 2-7*B* shows the resulting tool or gage configuration consisting of three toolmakers' buttons that establish datum reference plane *A*, the two tangent locator elements that establish datum reference plane *B*, and the datum reference plane *C* tangent locator element.

Figure 2-8*A* shows one way a drawing may differentiate between tangent part locators and toolmakers' buttons. The tool-locator symbols placed on (or straddling) the part contour in Fig. 2-7*A* and positioned in the adjacent view, usually indicate the position of these buttons which are shown in the tool or gage datum element configuration in Fig. 2-8*B*.

Figure 2-9*A* shows the alignment that could occur if part datum feature *A* contacts a tool make up of datum elements consisting of tangent and button locators. Figure 2-9*B* shows the identical part placed in a gage consisting of rail datum elements, and Fig. 2-9*C* shows this part will not align in the rail gage as it did in the tool. These setup errors can occur in all three datum planes and may cause the rejection of many parts because

14   *Fundamentals of Positional Tolerancing*   Chap. 2

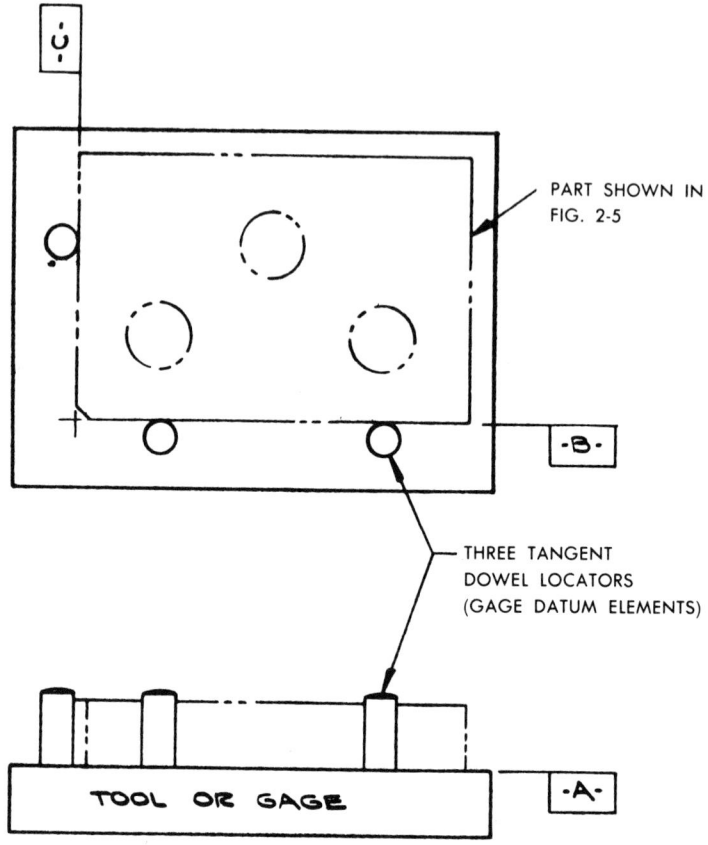

Fig. 2-6. Tool or gage tangent locators.

the same reference plane is not used in each case when the tool or gage design changes. Perfectly square and flat part datum surfaces would not result in any misalignment if different tools and gages were used, but since the perfect part does not exist, it is more practical to specify the actual datum reference planes used by defining the tool or gage datum element geometry. When the part datum features firmly contact the tool or gage datum elements, the actual "datum" for manufacturing and measurement is established. In surface plate inspection, the geometry of angle plates, close fitting pins, and the surface plate establish datum elements that, when contacted by part datum features, also determine the actual "datum" for measurement. The use of identical tool and gage elements will ensure that the least amount of tolerance is "robbed" from all features dimensioned from the part datum surfaces. Both the tool and the gage should simulate the mating part where they contact the part datum surface if at all possible.

Chap. 2 Fundamentals of Positional Tolerancing

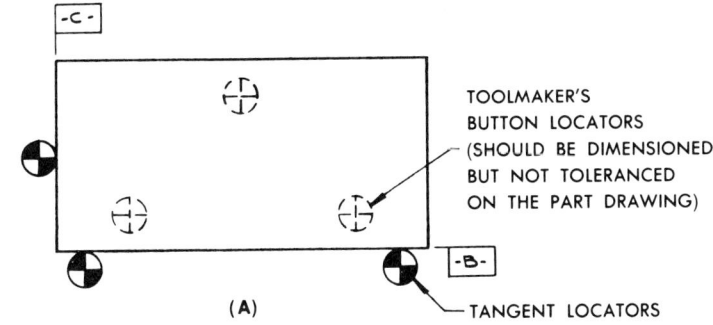

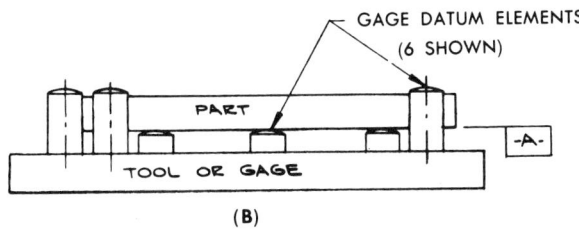

Fig. 2-7. Combined tangent and button locators.

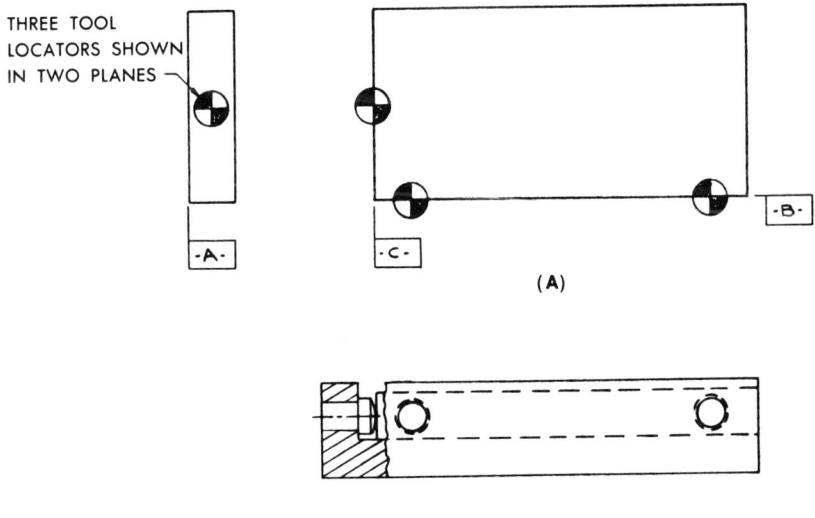

Fig. 2-8. Tool or gage locator symbols and the resulting tool or gage.

16   *Fundamentals of Positional Tolerancing*   Chap. 2

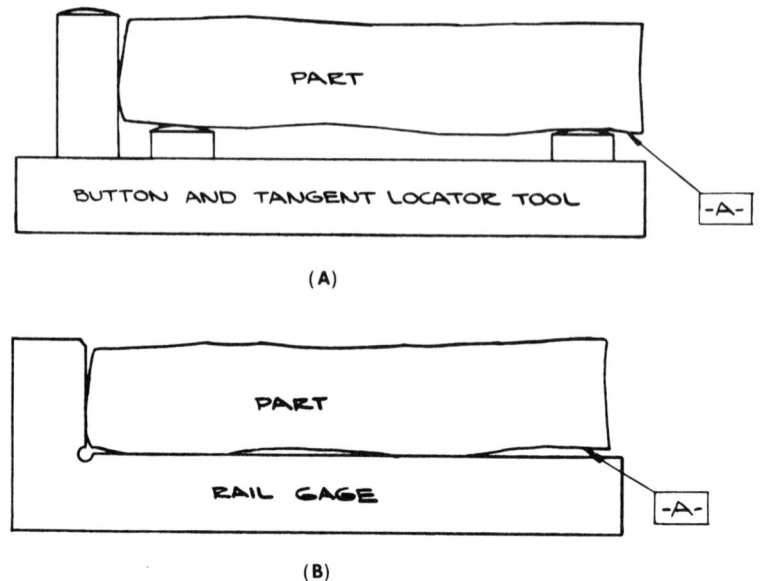

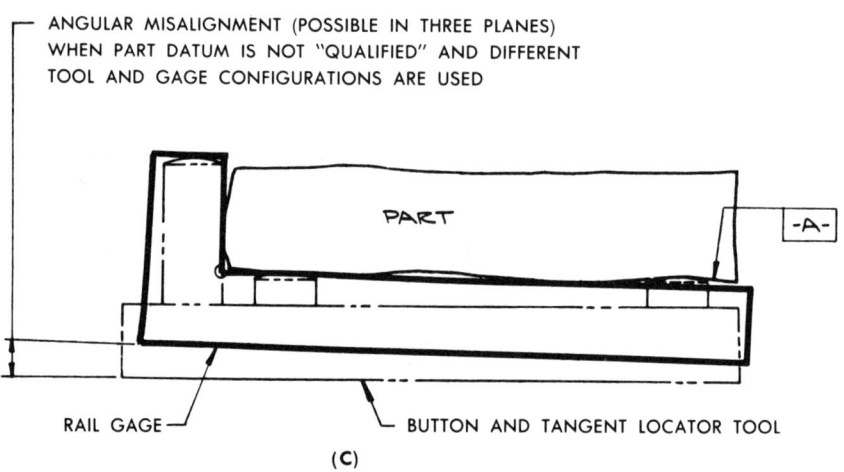

Fig. 2-9. Possible setup errors.

This simulation will reduce setup errors, since it is both practical and natural, and since it will deform the "datum" on flexible parts no more than the mating part at assembly.

To summarize, part features are dimensioned from part datum surfaces, but are measured from datum reference planes associated with part datum surfaces.

Chap. 2 *Fundamentals of Positional Tolerancing*   17

Positional Tolerance Principles

True position dimensions denote basic or theoretically exact (and therefore desired) relationships, locations, or sizes. All dimension lines drawn perpendicular to one another, and which establish the true position of surfaces, features, and feature patterns in relation to part datum surfaces, are perfect. True position dimensions are enclosed in rectangles and are not practical unless toleranced. Actually, in any dimensioning and tolerancing system, dimensions themselves do not have any tolerance and are considered perfect. Basic dimensions denote that tolerances will not accumulate even when basic dimensions are placed in a series or are "chain dimensioned." Chain dimensioning between part datum features does accumulate tolerances, however, and should be avoided unless it is obvious to the designer, because of large part size or configuration, that several different setups or tools are mandatory in the manufacture and measurement processes.

True position tolerances are the total permissible variation in the loca-

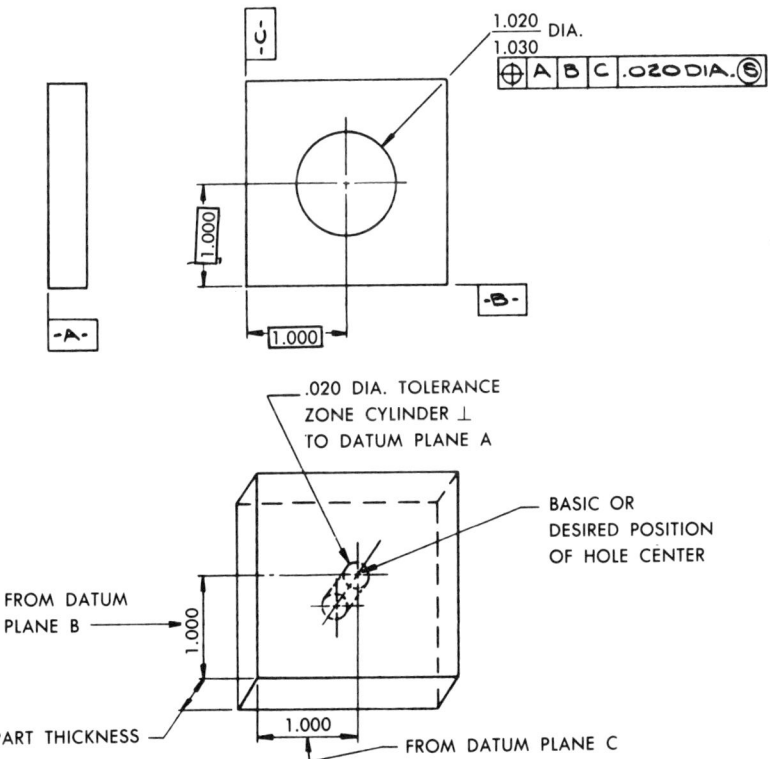

Fig. 2-10. Cylindrical tolerance zone.

tion of features and surfaces from their true position, size, or form. True position dimensions indicate not only the basic location of part features, but also the location of fixed gage elements that will in turn determine the relationships and/or locations of the part features.

True position tolerances are expressed as constant diameters or widths that are applicable throughout the entire depth, length, or circumference of parts unless otherwise specified. Figure 2-10 shows a cylindrical tolerance zone specified to a single hole located with basic dimensions. Unless otherwise specified, this zone includes both the positional and perpendicularity tolerance of the hole. This tolerance zone is perpendicular to primary part datum plane A.

Figure 2-11 shows a width or rectangular tolerance zone that includes slot squareness and location. This tolerance zone is perpendicular to part datum plane A and parallel to part datum plane B.

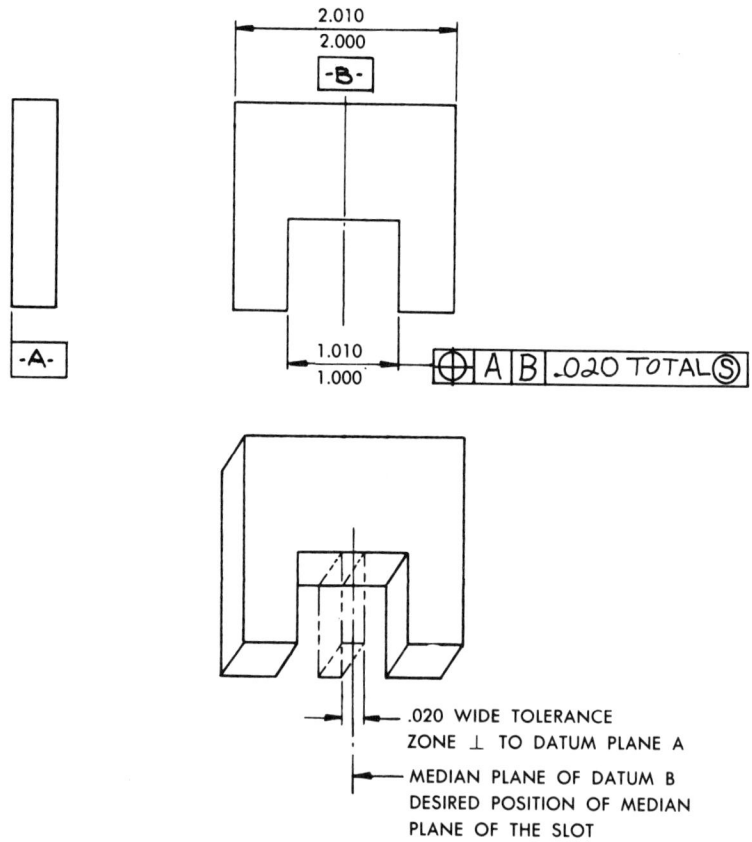

Fig. 2-11. Symmetry-type positional tolerance.

Chap. 2 *Fundamentals of Positional Tolerancing*

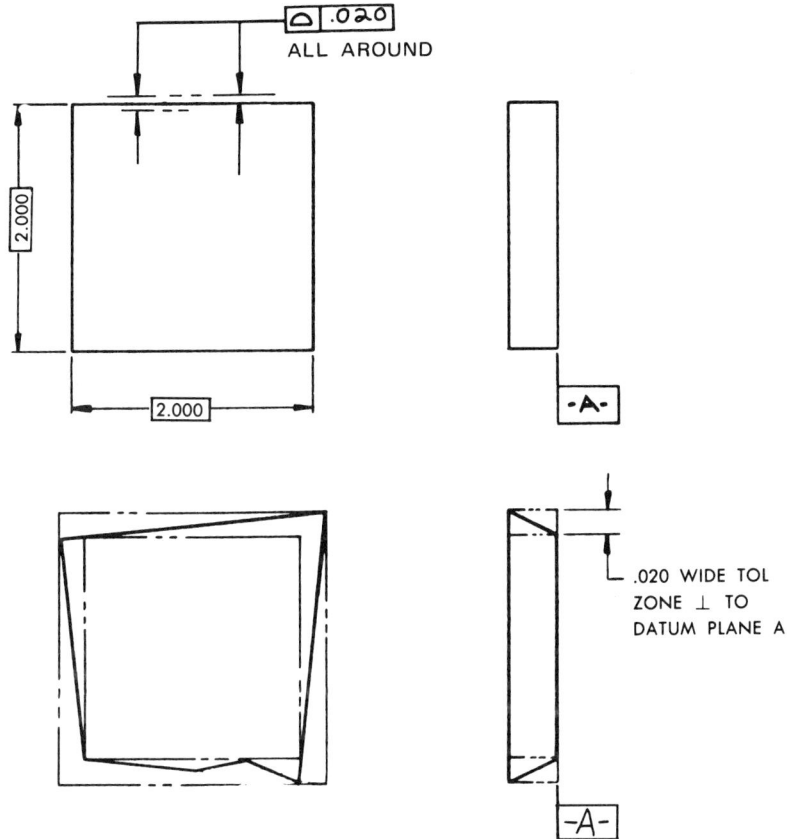

Fig. 2-12. Contour tolerance zone.

Figure 2-12 shows a tolerance zone that allows variation in part size, location, and form along a basic contour. This drawing shows the desired contour fully defined by basic dimensions and two phantom lines that indicate the size and location of the tolerance zone. For drawing clarity, the phantom lines will usually be drawn farther apart than actual scale size. The phantom line tolerance zone may be symmetrically located about the basic contour or may be entirely on one side or the other. However, there is no apparently practical reason to show the zone unequally disposed about the true contour since a true nominal part size will always technically exist halfway between the phantom tolerance zones. The contour tolerancing technique is compatible with optical inspecting methods that utilize optical projectors and chart gages.

Go and Not-Go Plug and Ring Gages—Taylor's Principle

The part features shown in this text will require plug or ring gages to check their minimum and maximum size. The Go plug or ring gage is not required, however, if a locational or form tolerance of 0.000-in. diam at MMC is specified, since the functional gage will include the Go gage element. Separate Go plug or ring gages are never needed to assure functional interchangeability in any case when fixed-element functional gages simulate the mating part at its MMC size.

Figure 2-13A shows a part with 1.000–1.010-in.-diam hole. The Go plug gage, on the left end of the gage in Fig. 2-13B, must be at least the maximum thickness of the part (or depth of the hole). The gage operator may lightly rotate or realign the Go gage to start it into the hole or over the shaft being inspected. The Not-Go gage, shown at the right end of the gage, need not be as long as the hole is deep. This gage should not enter the hole to a depth greater than one-third the hole depth, to allow for a slight bell mouth. The plug gages illustrated in Fig. 2-13 have not been toleranced; gage tolerances and allowances are discussed in MIL-STD-120 which states that up to 10 percent of the part tolerance, or 0.001 in. in this case,

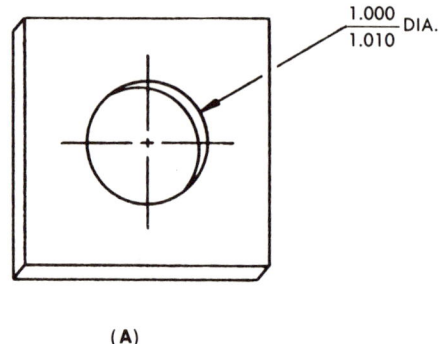

(A)

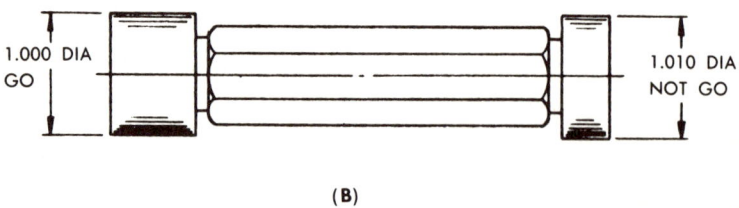

(B)

Fig. 2-13. Drawing and corresponding plug gage.

Chap. 2 Fundamentals of Positional Tolerancing 21

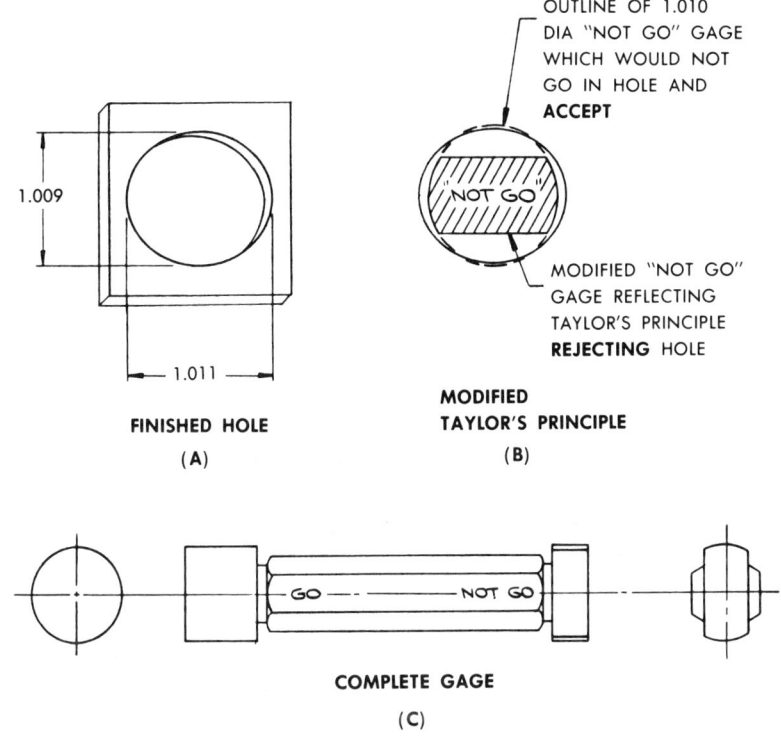

Fig. 2-14. Taylor's principle (modified).

can be used as the gage tolerance (5 percent) on both the Go and Not-Go gage elements. In addition, up to 5 percent more of the part tolerance can be used on the Go gage as a wear allowance.

Taylor's Principle. "The maximum material limits of as many related dimensions as possible or convenient should be incorporated in the Go gage; whereas the minimum material limits of these dimensions should be gaged by separate Not-Go gages."* Plug and ring gages that reflect the geometry of the workpiece feature they inspect do not, however, consistently accept "in-tolerance" parts. This comment pertains only to the Not-Go gage member. Figure 2-14A shows the finished part hole from Fig. 2-13A as elliptical and out of tolerance (1.011-in. diam). A standard 1.010-in.-diam Not-Go plug gage will not enter this hole.

Figure 2-14B shows the same 1.010-in.-diam Not-Go plug gage shown in Fig. 2-13B modified by a grinding operation so that when rotated it may

---
*Principles of Engineering Inspection, King and Butler, Cleaver-Hume Press, Ltd., London, 1957.

22   Fundamentals of Positional Tolerancing   Chap. 2

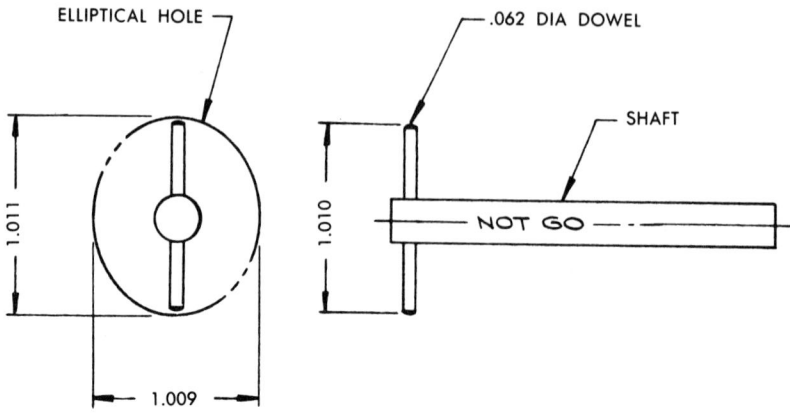

Fig. 2-15. Pin-type Not-Go gage for out-of-round holes (Taylor's principle).

enter such an elliptical hole when the hole is too large. Figure 2-14C shows the complete gage. The shape of this Not-Go plug gage should theoretically consist of only two points at each end of a diamond-shaped pin to completely fulfill its gaging function. A 0.062-in.-diam dowel pin would perhaps be adequate if used in a gage as shown in Fig. 2-15, providing that the shaft is parallel to the hole. If this gage could be inserted into a deep hole at a slight shaft angle and then be rocked radially, the hole would, of course, be out of tolerance because it is too large. A telescoping hole gage could be set at 1.010 in. to perform the same function.

Figure 2-16 shows Taylor's principle applied to the Not-Go gage for a shaft. Fortunately, most snap gages reflect this principle.

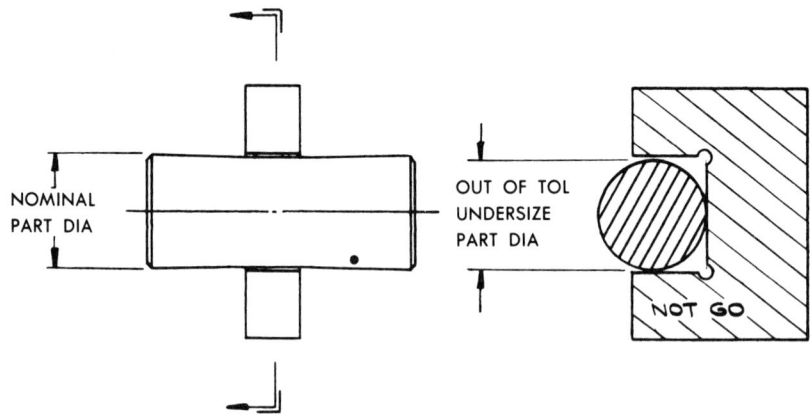

Fig. 2-16. Application of Taylor's principle to ring gage.

Chap. 2 Fundamentals of Positional Tolerancing

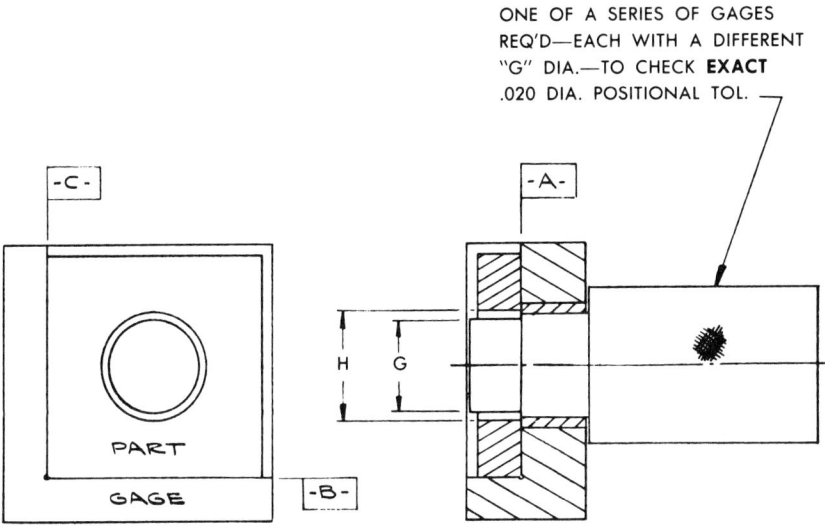

Fig. 2-17. Gage cited in Tables 2-1 and 2-2.

Regardless of Feature Size (RFS)

Tolerances modified with RFS (Regardless of Feature Size) callout or symbol (Fig. 2-10), are *constants* and cannot be increased. The RFS specification is not a practical gaging callout since a series of gages, all with different interchangeable G diameter pins (Fig. 2-17), would have to be made to check the finished parts when the finished holes varied in diameter between 1.020 in. and 1.030 in. Table 2-1 shows the different G gage pins required.

Table 2-1. RFS Gage Element Diameters

Finished hole diam H, in.	Gage element diam G, in.	RFS positional and perpendicularity tolerance, in.* (A non-variable tolerance)
1.020	1.000	0.020
1.021	1.001	0.020
1.022	1.002	0.020
1.023	1.003	0.020
1.024	1.004	0.020
1.030	1.010	0.020

*Difference in gage and hole size, since 0.020 in. is the maximum tolerance allowed.
Even a series of gages, as shown in Table 2-1, would not *exactly* meet the RFS specification unless they were *exactly* 0.020-in. diam smaller than the finished workpiece hole size. The RFS tolerance modifier would therefore actually require gage sizes as shown in Table 2-2. Thus, to gage this part with its RFS callout, an infinite range of gage elements (an impossible requirement) would be required. The RFS callout does not specify the use of *fixed*-size gaging elements which functionally represent the firm interchangeability requirements of most mating parts.

Table 2-2. Effect of RFS Modifier on Gage Sizes

Finished hole diam H, in.	Gage element diam G, in.	RFS positional and perpendicularity tolerance, in. (A non-variable tolerance)
1.023	1.003	0.020
1.0327	1.0127	0.0200
1.02981	1.00981	0.02000
etc. —	etc. —	etc. —

Maximum Material Condition (MMC)

Tolerances modified with the MMC (Maximum Material Condition) callout are variable tolerances and may increase as the part feature is finished away from its most critical interchangeability size. This specification is a practical gaging callout since a gaging element of fixed size and location will automatically allow part hole location and perpendicularity to vary as finished holes vary in size. Table 2-3 shows that a fixed-size gage element used to gage the finished part shown in Fig. 2-18 will allow varying tolerances. This fixed-gage-pin element could represent a firm design requirement: for instance, that a 1.000-in.-diam bolt must pass through the hole.

If Fig. 2-10 had specified MMC instead of RFS* (see Fig. 2-19A), positional tolerance could vary as shown in Table 2-3. Note that since the positional tolerance specified (0.020-in. diam) must be held only when the part hole is at MMC (1.020-in. diam), this tolerance may increase by the exact amount the hole departs from its most critical (MMC) size. The MMC specification, if substituted for the RFS callout shown in Fig. 2-10, requires the 1.000-in.-diam gage pin shown in Table 2-3 and illustrated in Fig. 2-18. This 1.000-in.-diam gage element is determined by subtracting

Table 2-3. Positional Tolerances at MMC

Finished hole diam H, in.	Fixed gage element pin diam G, in.	Positional and perpendicularity tolerance. A variable diam, in.
1.020	1.000	0.020
1.0215	1.000	0.0215
1.026	1.000	0.026
1.0272	1.000	0.0272
1.0291	1.000	0.0291

---

*ANS Y14.5 specifies that all positional tolerances are at MMC unless RFS is specified, and that all form tolerances are RFS unless MMC is specified.

Chap. 2 Fundamentals of Positional Tolerancing   25

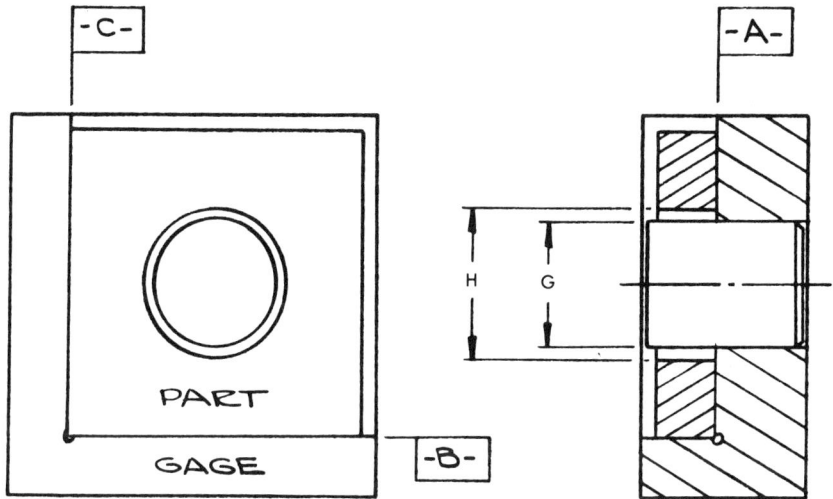

Fig. 2-18. Gage cited in Table 2-3.

the 0.020-in.-diam true-position tolerance zone specified at MMC from the MMC size of the hole, thus:

    1.020 in., MMC or most critical hole size
minus 0.020 in., true position tolerance allowed at MMC
    1.000 in., basic diam fixed gage feature

The following general rules should be followed in the use of MMC gages. (1) The gage pin should be the same form as the part feature and the true-position tolerance zone. (2) Subtract the positional tolerance from the MMC size of internal features to obtain basic gage element. (3) Conversely, add the positional tolerance to the MMC size of external features to obtain basic gage element size.

H minus *G* equals a variable tolerance when MMC is specified. The true functional requirement always *remains* firm and is represented by gage pin *G* in Fig. 2-18. Due to inherent process variations, parts will vary and no two part features will ever be exactly alike. This variation is represented by *H* in Fig. 2-18.

Since the 1.000-in. gage pin checks only hole location and perpendicularity at MMC, the size of the hole must be checked separately with 1.020-in.-diam Go and 1.030-in.-diam Not-Go gages. However, since the 1.000-in.-diam fixed gage pin simulates a 1.000-in.-diam bolt in the mating part, the use of a separate 1.020-in.-diam Go gage will reject the part if the hole is smaller than 1.020 in. *even if this hole is accepted by the 1.000-in.-diam fixed gage pin.* Thus, the very best parts (with holes *undersize* but located

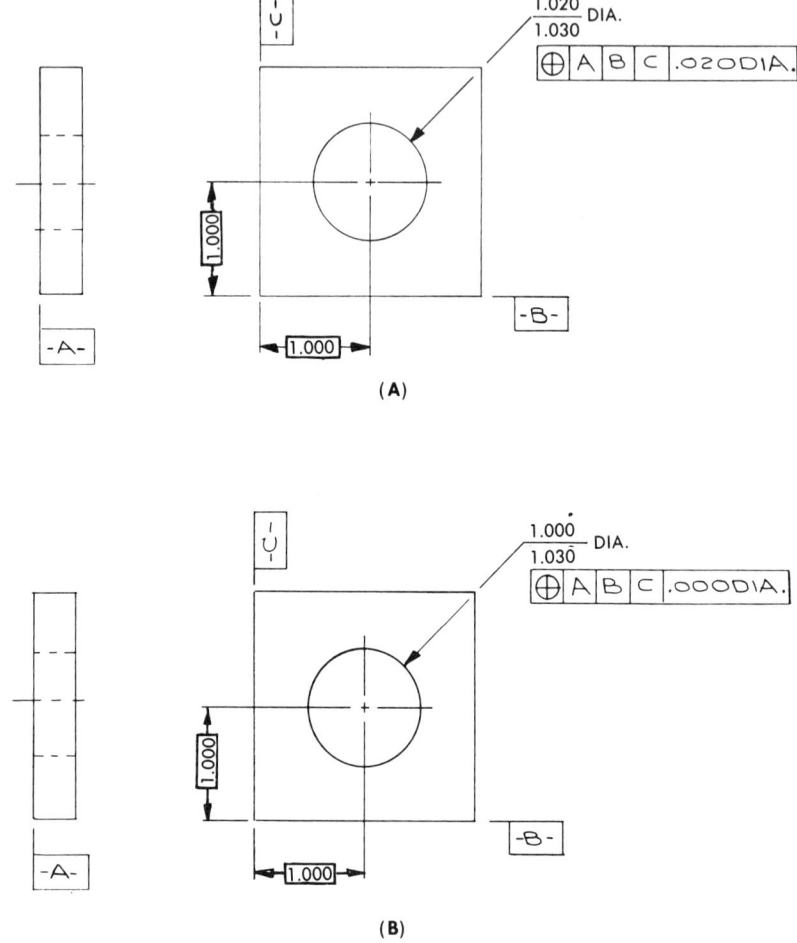

Fig. 2-19. Conventional and zero tolerancing.

more precisely than the 0.020-in.-diam positional tolerance specified) will be rejected. Millions of dollars are wasted each year because of the traditional use of such Go gages.

The separate 1.020-in.-diam Go gage can be eliminated and incorporated in the functional gage if the specification shown in Fig. 2-19*B* is used.

A callout that requires a "perfect" hole (0.000-in. diam) at MMC makes the positional tolerance *completely dependent* on hole size as shown in Table 2-4. The callout saves separate Go gage cost and operator time, and also allows the manufacturer the unheard-of freedom to pick his own working tolerance when he chooses a drill size.

Chap. 2   Fundamentals of Positional Tolerancing

Table 2-4.   Relation Between Positional Tolerance and Hole Size

Finished workpiece hole diam, in.	Positional and perpendicularity tolerance diam, in.
1.000	0.000
1.001	0.001
1.002	0.002
1.003	0.003
1.004	0.004
1.005	0.005
1.006	0.006
1.007	0.007
1.008	0.008
1.009	0.009
1.010	0.010
etc.	etc.

Basic Interchangeability Gages

The following gages will guarantee the interchangeability of the parts in Cases I, IIA, and IIB below when each part feature (hole, tapped hole, etc.) in each pattern of features has the same basic locations.

Case I: Clearance holes in mating parts

Rule: The gage for each part consists of a pattern of pins each centered at part basic hole locations. The gage pins will be the MMC size of the assembly screws or bolts (Fig. 2-20).

Case IIA: Clearance holes in part 1 (Fig. 2-20) and dowel pins or studs in part 2 (Fig. 2-21)

Rules: *Part 1*

The gage for part 1 (Fig. 2-20) consists of a pattern of pins each centered at part basic hole locations. The gage pins will be the MMC size of the part 2 dowel or stud plus the positional toler-

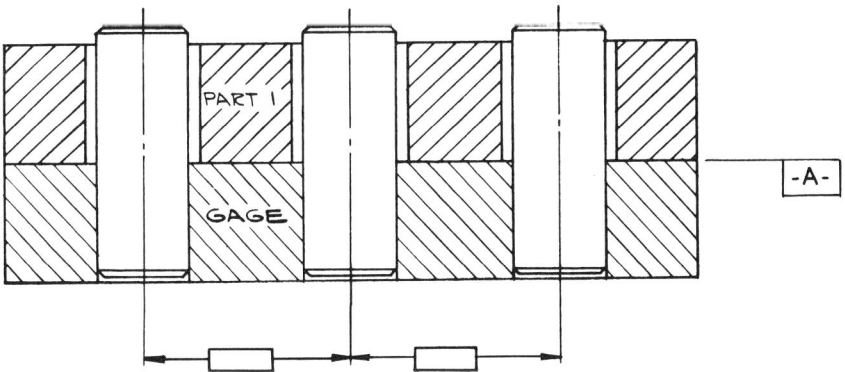

Fig. 2-20.   Gage for part 1 (clearance holes).

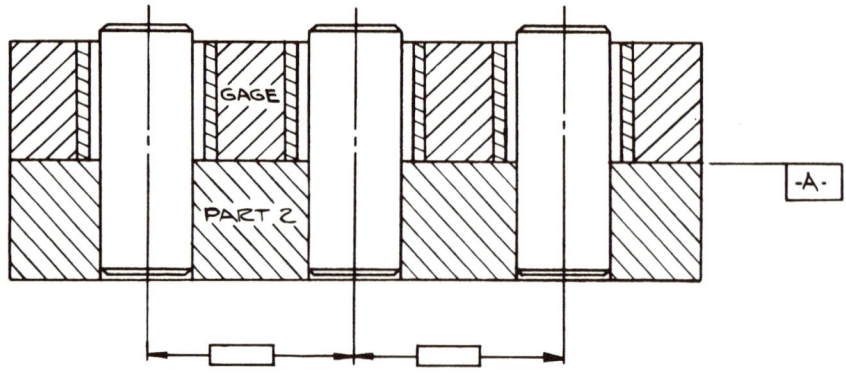

Fig. 2-21. Gage for part 2 (studs or dowels).

ance diameter specified at MMC for the dowels or studs. The positional tolerance increases the virtual size of dowels or studs, and makes them actually larger at assembly.

Part 2

The gage for part 2 (Fig. 2-21) consists of a pattern of bushings each centered at part basic dowel or stud locations. The bushing ID's will be the part 2 dowel or stud MMC size plus the positional tolerance diameter specified at MMC for the dowels or studs.

Case IIB: Clearance hole in part 1 (Fig. 2-20)
Tapped holes in part 3 (Fig. 2-22)

Rules: *Part 1*

The gage for part 1 consists of a pattern of pins each centered at part basic hole locations. The gage pins will be the part 3 tapped thread size plus the positional tolerance diameter specified at MMC for the tapped features.

Part 3

The gage for part 3 (Fig. 2-22) consists of a pattern of bushings each centered at part basic tapped hole locations and a series of Go thread gages, one for each tapped hole in the pattern. The difference between the bushing ID (diam B) and the shank diameter of the Go thread diameter where it goes through the gage bushing (diam G) will be the positional tolerance specified at MMC for the tapped thread. The bushing length will equal the thickness of the mating part.

Chap. 2 Fundamentals of Positional Tolerancing 29

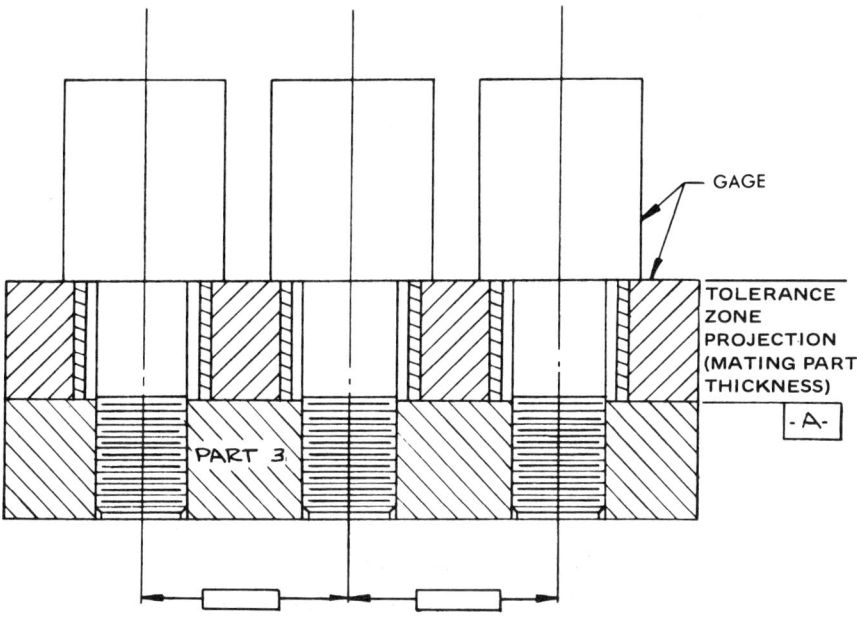

Fig. 2-22. Gage for part 3 (tapped hole).

CHAPTER 3

DESIGN PRINCIPLES FOR FEATURE RELATION GAGING

The gage designs discussed in this chapter apply to all patterns containing two or more features, provided none of the features is designated as a part datum feature. Only feature relation gages are discussed; combined feature-location-and-relation gages will be covered in Chapter 4. All the gages discussed in this chapter must contact the primary part datum surface (A) during use, and suitable mechanical or optical means must be used to insure contact.*

INTERNAL FEATURE PATTERNS

Clearance Hole Patterns

Figure 3-1A shows a six-hole pattern of 0.510–0.530-in.-diam clearance holes in which each hole has a positional tolerance of 0.010-in.-diam when the hole is at MMC. Note that only one primary datum feature A on the part is specified, which indicates that only a feature relation gage is required. This gage will check the hole pattern relationship and perpendicularity to datum plane A, but not the location of the pattern on the part.

---

*Fig. 4-17 illustrates the most practical method of dimensioning and gaging the location of the features on these gages.

Chap. 3 Design Principles for Feature Relation Gaging

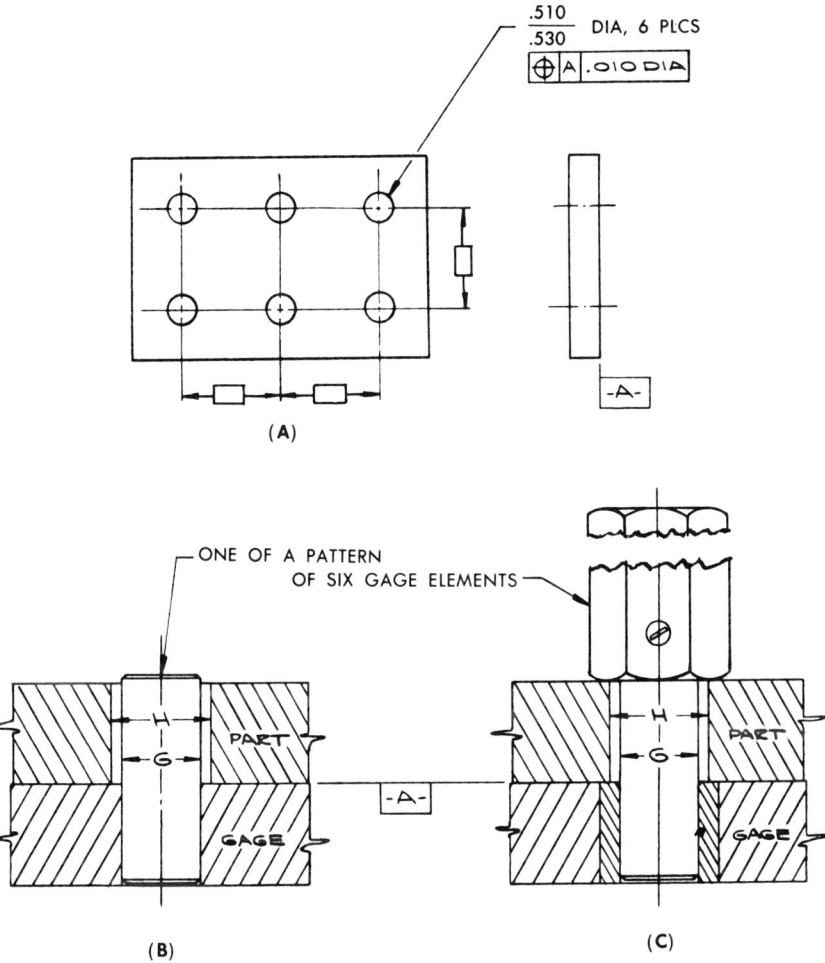

Fig. 3-1. Part with clearance holes and functional gage elements.

Functional Gage Design. The gage element locations are identical to the basic dimensions on the part drawing (both patterns are identical), and the gage element pin G (Fig. 3-1B or C) must have a minimum height that is at least the maximum thickness of the mating part. This gage element pin will check hole location and perpendicularity to datum plane A, and the six gage element pins will be 0.500-in.-diam determined as follows:

$$\begin{array}{r}0.510\text{-in. hole }H\text{ diam at MMC}\\ \text{minus }0.010\text{-in. positional tolerance specified at MMC}\\ \hline 0.500\text{-in. diam basic size gage pin }G\end{array}$$

All six gage element pins must enter the six part features at the same time to fulfill Taylor's Principle (see p. 21) and the gage datum element must contact part datum surface A for acceptance. Thus: H minus G equals 0.010 in. at MMC.

Plug Gages Required. There will be required a 0.510-in.-diam Go gage, and a 0.530-in.-diam Not-Go gage.

If the drawing callout required $\begin{smallmatrix}.500\\.530\end{smallmatrix}$ diam (6 plcs) with positional tolerance 0.000-in.-diam at MMC, the identical 0.500-in.-diam gage elements shown in Fig. 3-1B or C would still be required (0.500-in.-diam hole H minus 0.000-in.-diam positional tolerance equals 0.500-in. gage element G). The 0.510-in.-diam Go plug gage would not be required, however, since it is built into the functional gage in the form of the six 0.500-in.-diam gage elements that simulate the six 0.500-in.-diam bolts that will fasten this part to the mating part. Figure 3-1C illustrates another functional gage with separate gage elements G that are inserted into the part holes and then into close-fitting (0.0001–0.0002-in.-allowance) bushings in the gage base. The same basic rules apply to this design as to the gage shown in Fig. 3-1B and all six gage elements must enter the bushing and be in place in order to accept the six-hole pattern for perpendicularity and feature relationship. Specific applications of the gage elements illustrated in Fig. 3-1B or C will be shown in subsequent examples, but it can be generally stated that the separate gage elements shown in Fig. 3-1C will enable the gage operator to determine which part feature is out of tolerance. All six gage elements must enter the six holes in the part and be in place simultaneously. The number of gaging elements should never be reduced for checking any feature pattern if separate gage elements are used (as shown in Fig. 3-1C) by "pinning" several holes and then "walking" *one* gage element around the part pattern. The part-gage relationship could easily shift during this type of gaging operation, and out-of-tolerance parts accepted. The entire *basic* feature pattern must be gaged simultaneously, since all bolts must be inserted as a pattern at assembly.

Tapped Feature Patterns

Figure 3-2A shows a six-feature pattern of tapped holes. Figure 3-2B shows the basic design of the gage that inspects each tapped hole in the pattern. The gage element G that goes through bushing B in the gage base and enters the tapped hole in the part simulates the bolt, and all six of these gage elements must be in place at the same time.

Functional Gage Design. The bushing locations on the gage base are identical to the basic dimensions on the part drawing (both patterns are identical), and the gage bushings have a minimum height that is at least the maximum thickness P of the mating part (0.50 in. in this example). The difference between gage base bushing diam B and that portion of the Go

Chap. 3   *Design Principles for Feature Relation Gaging*   33

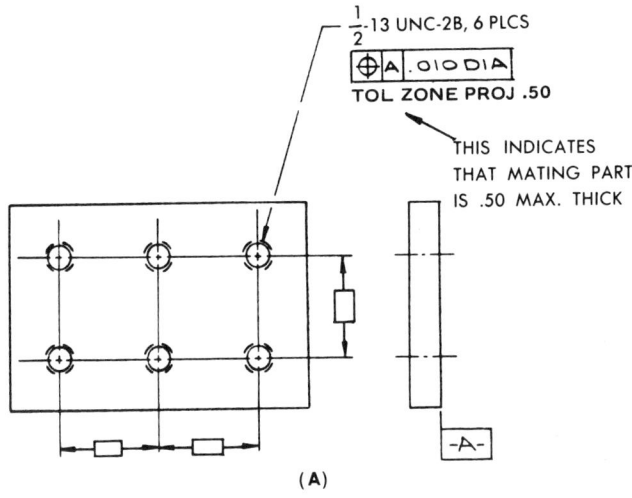

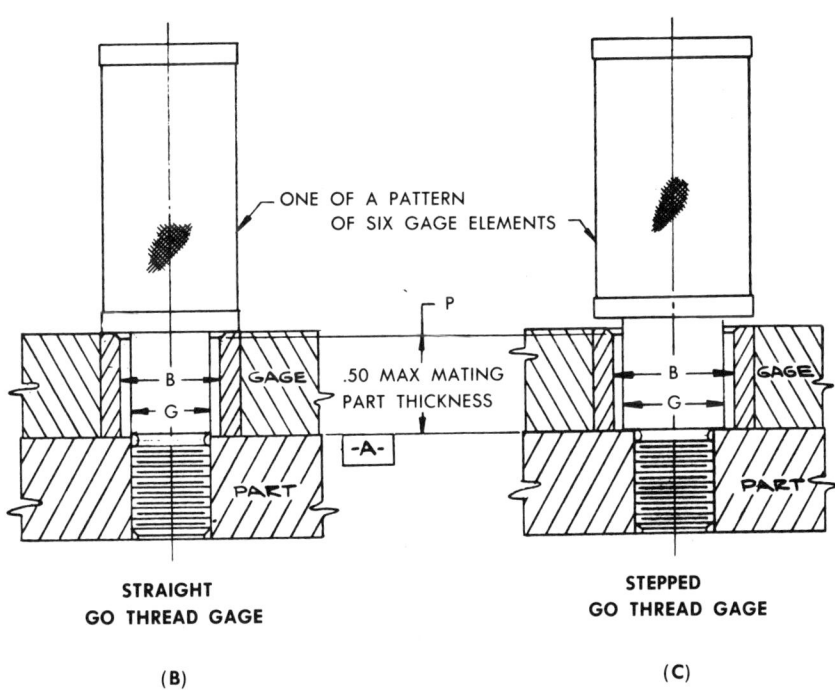

Fig. 3-2.   Part with tapped holes and functional gage elements.

thread gage element G (that simulates a bolt where it goes through the gage bushing) is the positional tolerance specified at MMC for the tapped features. Thus: B minus G equals 0.010-in. in Fig. 3-2B or 3-2C.

Plug Gages Required. Gage element G is a Go thread gage, and so only a Not-Go thread gage (not shown) is required to check each hole separately. Figure 3-2C shows the same basic gage design as Fig. 3-2B except that the Go thread gage member G has been "stepped" so that a standard size ID bushing B may be used. In Fig. 3-2B bushing B would have to be enlarged to obtain the 0.010-in. difference between the 0.500-in. G gage element, since a 0.510-in.-diam bushing is not standard.

Miscellaneous Considerations. The same gage would check a "Helicoil" or "Rosan" type threaded insert, but threaded inserts with locking features should not be checked since the Go thread element could damage the locking feature. It is better to gage the tapped hole pattern with a special in-process gage before installing the insert.

Because of tolerance there can be from 0.000 to 0.004-in. shake or movement between the pitch diameters of the Go thread gage elements and part tapped features, so the Maximum Material Condition modifier is applicable. If the positional tolerance for the tapped hole were modified with "Regardless of Feature Size," the Go thread gage element would have to *center* in the tapped thread. This would require either a tapered Go thread element or an expanding split thread arrangement. Thus, since very few mating parts have centering threads, the RFS callout is usually not practical.

Tapped features are gaged for relation and perpendicularity where the assembled bolt (simulated with the Go thread gage element) goes through the mating part feature (simulated with the gage bushing). The tapped thread is not checked for relation or perpendicularity *inside* the thread. Since the mating part thickness is quite critical in determining the allowable perpendicularity of the thread, this information is vital to the gage designer in determining gage bushing height, and should be specified on the part drawing.

Counterbore Patterns, Case I

Figure 3-3A shows a part with clearance holes and counterbored holes. Figure 3-3B shows the basic design of the gage that inspects a pattern of counterbores and holes when both counterbores and holes have the same specified positional tolerance, i.e., each counterbore is *not dimensioned from its respective clearance hole*.

The bushing locations on the gage base are identical to the basic dimensions on the part drawing (both patterns are identical). The combined counterbore-clearance hole gage element must fully enter both counterbore and gage base bushing, and bottom on the counterbore. The difference between part counterbore C' and gage element C, and part clearance hole H' and gage element H, equals the 0.010-in.-diam positional tolerance

Chap. 3 Design Principles for Feature Relation Gaging   35

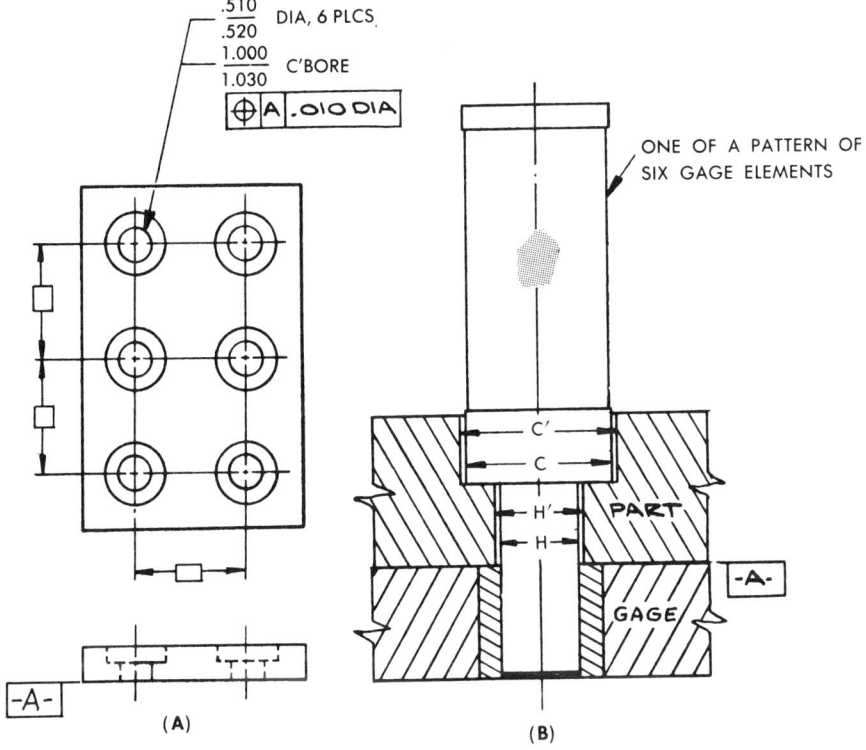

Fig. 3-3. Part with counterbores (Case I) and functional gage.

specified when both counterbores and clearance holes are at their MMC size. Thus:

0.510-in. hole H' diam at MMC
minus 0.010-in. positional tolerance specified at MMC
$\overline{0.500}$-in. = diam gage element H

Also, the 1.000-in. counterbore C' diam at MMC
minus the 0.010-in. positional tolerance specified at MMC
$\overline{0.990}$-in. = diam gage element C

Plug Gages Required. The following gages are required: (1) 0.510-in.-diam Go gage, (2) 0.520-in.-diam Not-Go gage, (3) 1.000-in.-diam Go gage, and (4) 1.030-in.-diam Not-Go gage.

Counterbore Patterns, Case II

Figure 3-4A shows a second method of specifying clearance and counterbored holes when each counterbore is dimensioned from its respective

36 Design Principles for Feature Relation Gaging Chap. 3

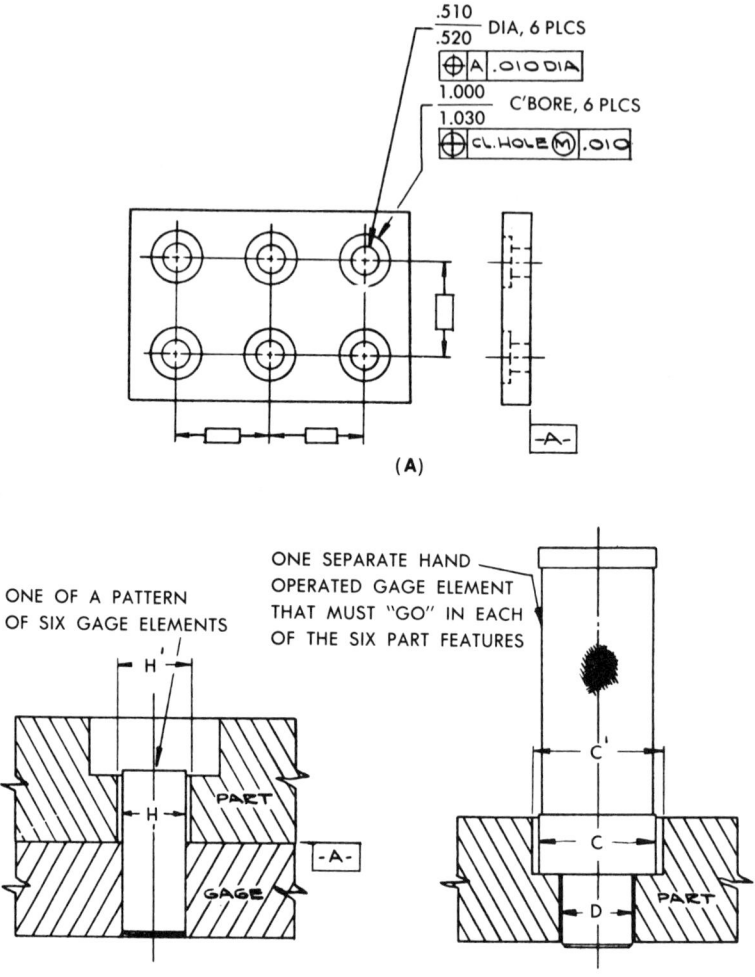

Fig. 3-4. Part with counterbores (Case II) and functional gage.

clearance hole. Figures 3-4*B* and 3-4*C* illustrate the two gages (and operations) required when the clearance hole pattern is dimensioned and toleranced separately from the counterbore pattern; i.e., each counterbore is located from its respective clearance hole. Figure 3-4*B* shows the gage that checks the positional tolerance of the clearance hole pattern. The size of these fixed gage pins is determined by subtracting 0.010-in. positional tolerance at MMC from the 0.510-in. MMC size of the hole. Thus:

Chap. 3 Design Principles for Feature Relation Gaging

0.510 in., MMC size of clearance hole
minus 0.010 in., positional tolerance specified at MMC
0.500 in., fixed gage pin size H
($H' - H = 0.010$ in.)

Figure 3-4C shows the separate hand-held gage that checks each counterbore from its respective clearance hole at MMC. Rule:
1. Gage size D is the MMC size of the clearance hole: thus, $D = 0.510$ in.
2. Determine gage size C by subtracting the 0.010-in.-diam positional tolerance at MMC from the MMC size of the counterbore; thus, 1.00 in. $-$ 0.010 in. $=$ 0.990 in. ($C' - C = 0.010$ in.).

All gage pin elements must simultaneously go into the clearance holes shown in Fig. 3-4B and each counterbore must be inspected separately, using the hand-held gage element shown in Fig. 3-4C. The 0.510-in.-diam datum element on the hand-held gage in Fig. 3-4C is a Go gage, so only the following plug gages are required: (1) 0.520-in.-diam "Not-Go," (2) 1.000-in.-diam "Go," and (3) 1.030-in.-diam "Not-Go."

Fixed-Nut Retainer Patterns

Figure 3-5A shows a pattern of fixed-nut retainers. Figure 3-5B shows the basic design of a gage that inspects a part containing a pattern of fixed-

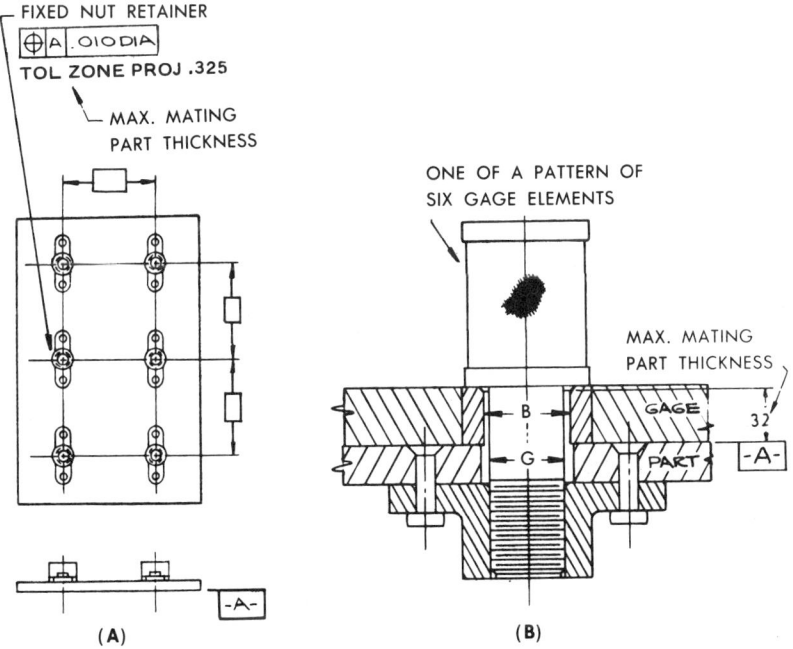

Fig. 3-5. Part with fixed-nut retainers, and gage.

nut retainers (i.e., the nut is rigidly held in the retainer).

With such a pattern, only the location of the thread in the fixed nut in the retainer must be gaged. *The access hole in the part need not be gaged for location.* As long as the access hole allows gage element G (which simulates the bolt) to enter the fixed nut, it is large enough and properly located. The minimum gage bushing height (0.325 in.) shown in Fig. 3-5A is the maximum part thickness. The difference between gage base bushing diameter B and that portion of gage element diameter G (that simulates the bolt) is the positional tolerance specified for the fixed nut in the nut retainer. Thus $B - G = 0.010$ inch.

This gage is similar to the gage in Fig. 3-2B since the fixed nut retainer simply allows the designer to place a nut in an inaccessible location or on a part that is too thin to be tapped. The Go thread gage B element can be stepped, as previously shown in Fig. 3-2C. Care must be exercised in designing the amount of "step," however, to be sure that the Go thread gage element does not contact the access hole in the part.

Floating-Nut Retainer Patterns

Figures 3-6B and 3-6C illustrate the individual gages required to inspect the floating-nut retainers shown in Fig. 3-6A. The pattern relationship of the clearance holes in the part (not the floating nuts) is gaged with the gage shown in Fig. 3-6B. This gage is a pattern of *fixed* gage elements that must simultaneously enter their respective clearance holes. The fixed gage pins are located at the basic clearance hole locations shown on the part drawing. Thus: $H - G = 0.030$ in., the positional tolerance at MMC. The separate hand-held gage shown in Fig. 3-6C is inserted into each floating nut to determine if the nut has sufficient "float" to allow this hand-held gage (which simulates the bolt) to contact the clearance hole all around its circumference.

The mating part thickness is not specified in Fig. 3-6A because the floating nut in the nut retainer can be tilted and thus adjusted for misalignment when the bolt (simulated with the hand-held gage element in Fig. 3-6C) is inserted. The fixed gage elements mentioned above could fit into bushings in the gage base and be sequentially inserted into the part clearance holes until all are assembled (Fig. 3-6C).

EXTERNAL FEATURE PATTERNS

These gaging principles apply to all patterns containing two or more features, provided that none of the features is designated as a part datum-feature.

Stud Patterns

Figure 3-7A shows a six-stud pattern consisting of 0.500–0.499-in.-diam studs welded on a plate. The weld flash has been removed. Figure 3-7B

Chap. 3   Design Principles for Feature Relation Gaging   39

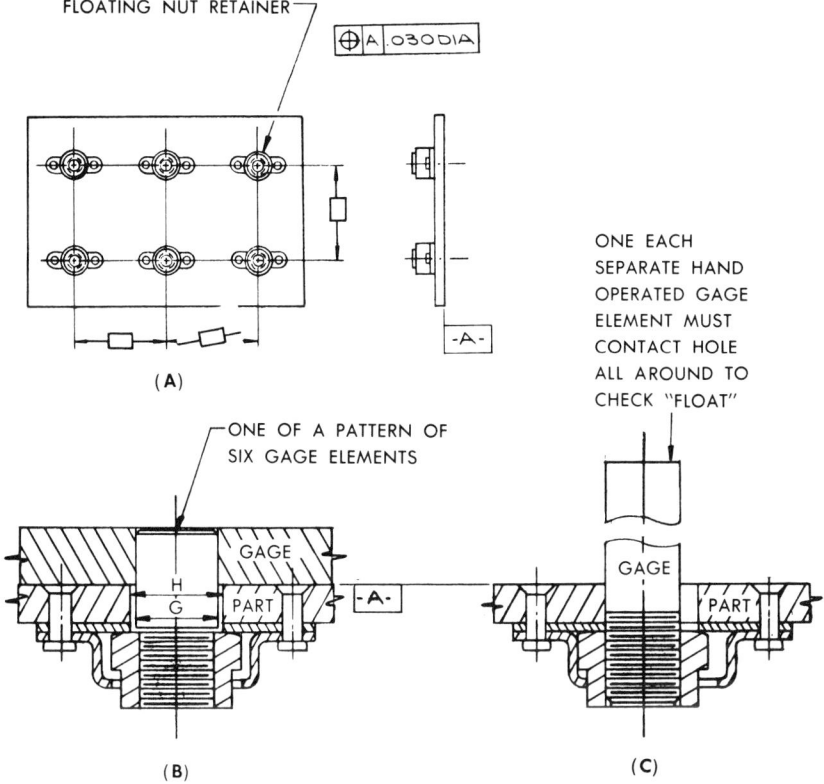

Fig. 3-6. Part with floating-nut retainers and functional gages.

shows the gage, and details one of six bushings, located at the same basic dimension as the studs in Fig. 3-7A. The bushing ID is determined as follows:

$\phantom{\text{plus }}$0.500 in., the specified MMC diam of the stud
plus 0.010 in., the positional tolerance specified at MMC
$\phantom{\text{plus }}$$\overline{0.510}$ in., bushing ID B

Thus, $B - S$ = 0.010 in. at MMC

For acceptance, all six bushings must go over the six studs, and the gage datum element must contact part datum feature A to accept the part. The ring gages required are a Go gage of 0.500-in. diam and a Not-Go gage of 0.499-in. diam.

Dowel Pin Patterns

Figure 3-8 shows a pattern of six 0.5000–0.4995-in.-diam dowel pins in a plate. Figure 3-7C shows the gage, and details one of six bushings,

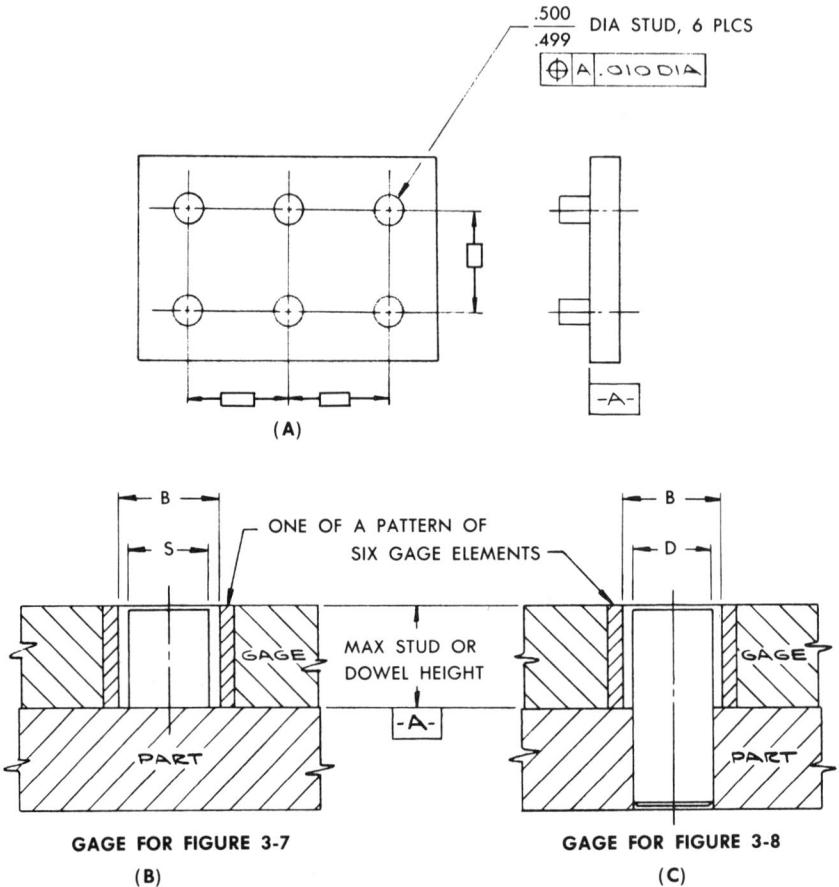

Fig. 3-7. Part with studs, and gages.

located at the same basic dimension as the dowels in Fig. 3-8. This gage is identical to the gage shown in Fig. 3-7B (except perhaps for bushing tolerance), and the bushing ID is determined as follows:

> 0.5000 in., the specified MMC diam of the dowel pin
> plus 0.0100 in., the positional tolerance specified at MMC
> 0.5100 in., bushing ID B

Thus: $B - D = 0.0100$ in. at MMC

All six bushings must go over the six dowels, and the gage datum element must contact part datum-feature A to accept the part. The ring gages required are a Go gage of 0.5000-in. diam and a Not-Go gage of 0.4995-in. diam.

Chap. 3   Design Principles for Feature Relation Gaging

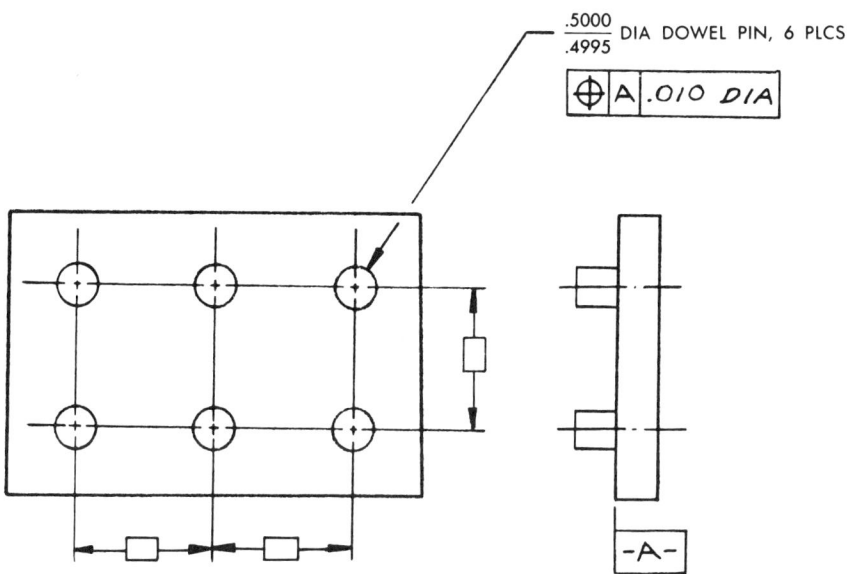

Fig. 3-8.   Part with dowel pins.

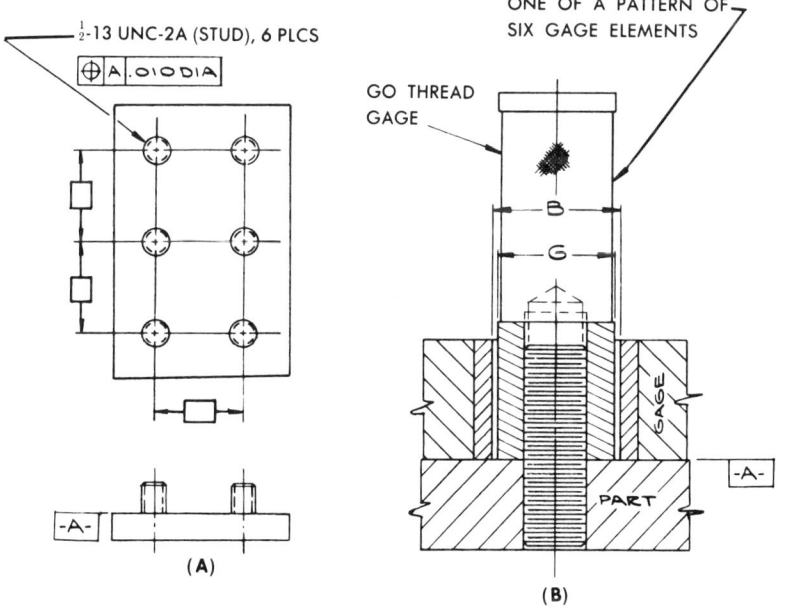

Fig. 3-9.   Part with threaded studs and functional gage.

Threaded Stud Patterns

Figure 3-9A shows a pattern of six threaded studs, and Fig. 3-9B shows the gage (one of six gaging elements) that gages each stud in the six-stud pattern. The difference between the OD G of the internal Go thread gage and the ID B of the gage bushing will be 0.010 inch. MMC has been specified for the location of the threaded studs and this is, in effect, the "shake" possible between the Go thread gage and the stud threads. There can be from 0.000 to 0.004-in. "shake" in the Go thread gage in many cases, so the MMC modifier is applicable. If the positional tolerance for the tapped stud were modified with RFS, the thread gage element would have to *center* on the stud, and this would demand an adjustable split Go thread element. If the positional tolerance were specified to the major diameter of the thread, the gage would consist of six fixed 0.510 ID bushings.

In this instance, only positional and perpendicularity tolerances are being checked, on the assumption that thread geometry has been previously checked and is acceptable.

CHAPTER 4

DESIGN PRINCIPLES FOR FEATURE LOCATION AND RELATION GAGING

Critical (RFS) Part Datum Features

All of the gages discussed in this chapter must contact the designated part datum features during use, and suitable mechanical means must be used to insure contact.

Part datum features modified with RFS require gage-centering devices so the gage will center on the part datum feature, regardless of its finished size. The datum feature must, of course, be in tolerance. Figure 4-1A shows a part datum feature modified with the RFS symbol and indicates, by the use of four tool and gage pickup symbols (or datum targets), where the part will be "centered."

Figure 4-1B shows the gage that will meet this drawing requirement. The four dial indicators located adjacent to the four holes in the pattern are specified per the tool and gage pickup symbols in Fig. 4-1A. The gage will have to be set with a master so that the indicators when zeroed (or so that the diametrically opposed indicators have the same readings) will center the part when it is placed in the gage. Note that all eight holes are checked simultaneously *as a pattern* with eight fixed gage elements which are located at the basic drawing dimensions. If the part will assemble on the eight gage elements when contacting the gage on datum surface A, and then can be centered in the gage by zeroing out the indicators, the part will

44 Design Principles for Feature Location and Relation Gaging   Chap. 4

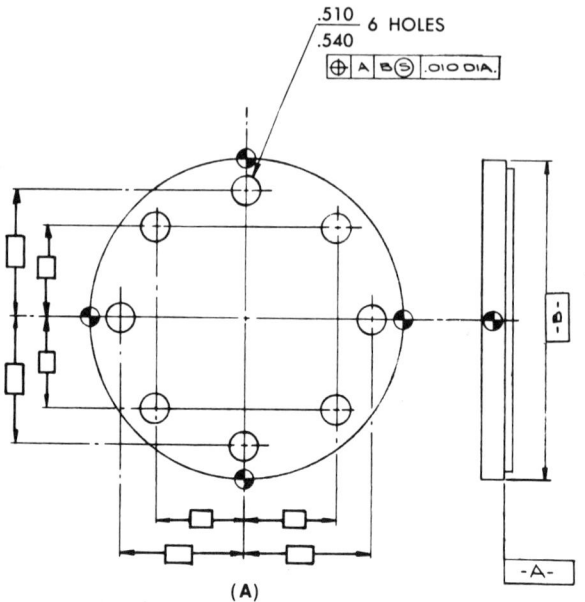

Fig. 4-1A. A critical (RFS) datum-feature, with tool and gage pickup points.

be accepted. Thus, the eight holes were located within their respective positional tolerances from that "center" specified on the part.

Fixed Gage Elements. Each of the eight gage pins will be 0.500-in. diam, determined as follows:

0.510 in., the MMC diameter of the part holes
minus 0.010 in., the positional tolerance specified at MMC
0.500 in., basic gage pin size

A 0.510-in. Go plug gage and 0.540-in. Not-Go plug are required.

Figure 4-2A shows a small central part datum feature modified with the RFS symbol. This drawing also indicates, by reference to both datum planes A and A', that the part is to be gaged twice in each of the gages shown in Figs. 4-2B and 4-2C.

Two steps are required in using the gage in Fig. 4-2B. First, the part must pass freely over the 0.406-in. Go gage pin, and part datum surface A must then fit flat against the gage datum element. Second, the part must also pass freely over the 0.406-in. Go gage pin, and part datum surface A' must then fit flat against the gage datum element. These two steps check that the 0.406–0.408-in.-diam part datum feature is perpendicular to both datum planes at MMC.

Chap. 4   Design Principles for Feature Location and Relation Gaging   45

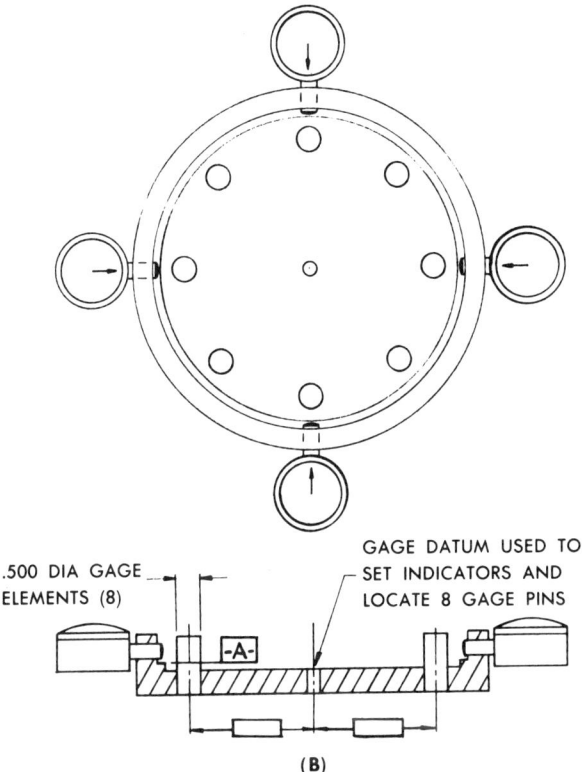

Fig. 4-1B.   Functional gage (with dial indicators) for a critical (RFS) datum-feature.

Four steps are required in using the gage in Fig. 4-2C, the same steps being repeated for the two datum planes, A and A'. (1) Locate part datum feature A on the gage base datum plane A; (2) insert datum gage element with diameters D (a 0.406-in. diam) and D' (somewhat larger than 0.408 in.), and center the part datum hole on the taper between D and D'; (3) insert the G gage elements and rotate the part as necessary to effect their entry; and (4) position all four G gage elements at the same time to accept the part.

For this operation, three plug gages are required: a 0.380-in.-diam Go gage, a 0.388-in.-diam Not-Go gage, and a 0.408-in.-diam Not-Go gage for the "datum" hole. The gage element in Fig. 4-2B is the 0.406-in.-diam Go gage.

Part datum features modified with the symbol RFS are expensive to gage and should be encountered only rarely when *mating part* datum-features

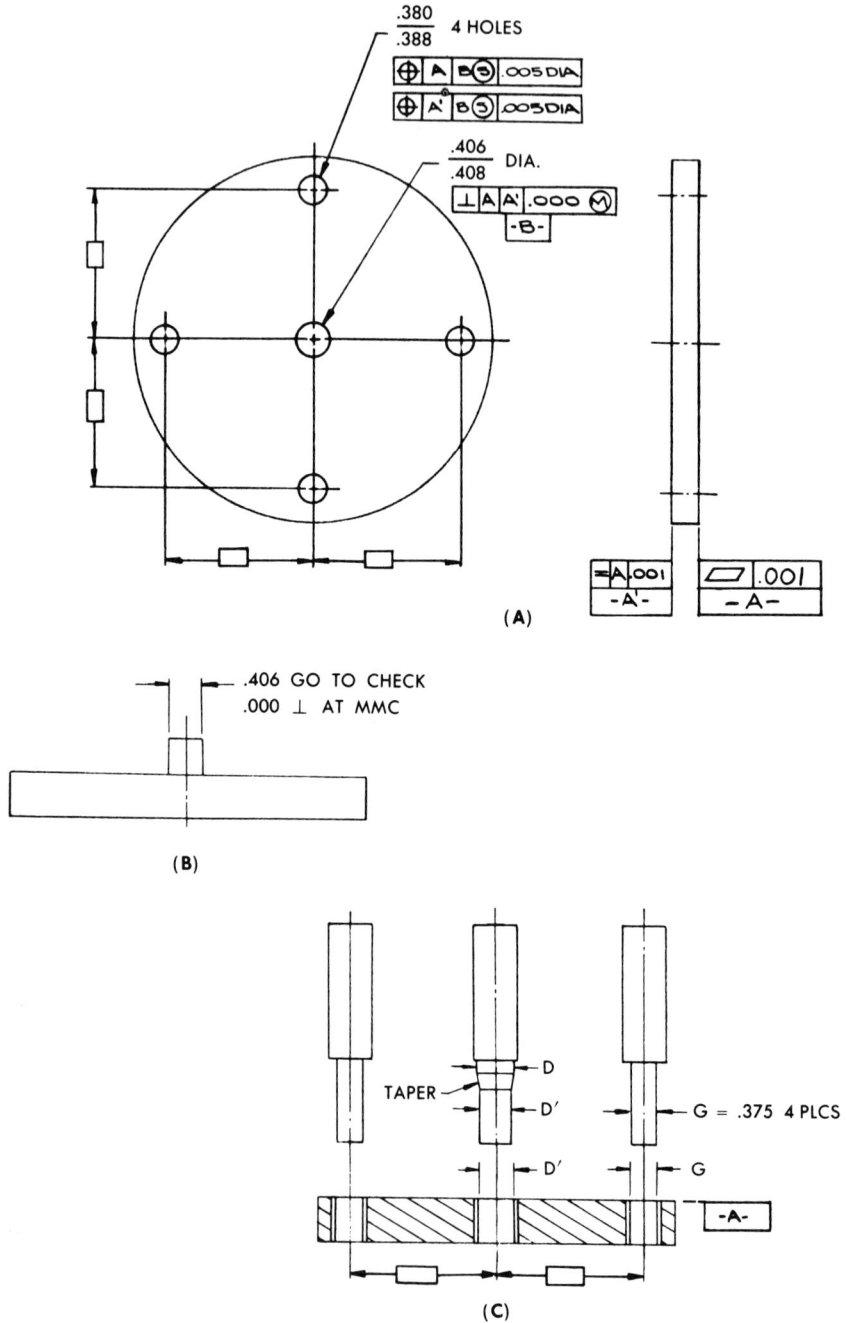

Fig. 4-2. A critical (RFS) datum-feature, with two required gages.

Chap. 4   Design Principles for Feature Location and Relation Gaging   47

consist of a tapered centered device similar to that shown in Fig. 4-2C. It is very difficult to repeatedly pick up the same "center" of a part datum feature unless that feature has perfect form.

Figure 4-3A shows tapered part datum-feature A modified with the RFS specification symbol. The clearance and counterbored holes and the two tapped holes are all dimensioned from the "center axis" of the taper. The mating part that contains clearance holes for the two screws that enter the tapped holes has a thickness of 1.00 inch. This is indicated by the 1.00-in. projected tolerance zone. Figure 4-3B shows a gage that verifies this part requirement. The tapered gage datum element should be the nominal size of the part datum feature.

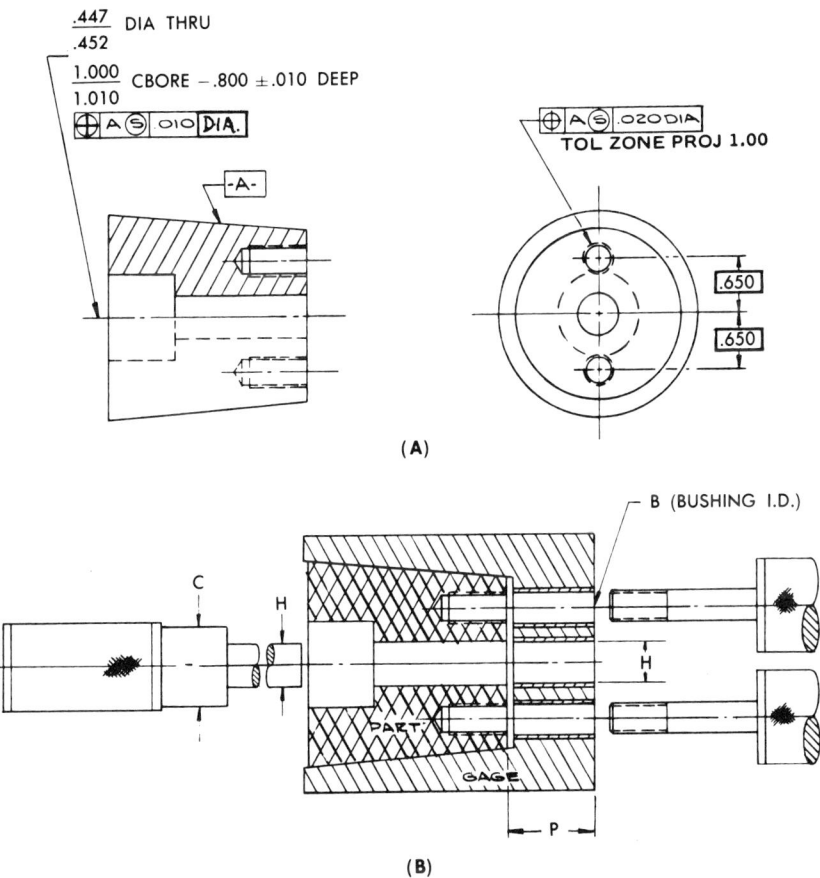

Fig. 4-3.   A critical (RFS) tapered-part datum-feature and functional gage.

The gage should be used as follows:

1. Insert the part into the gage and rotate if necessary so that the two Go thread gage elements will enter the threads, and then screw them partially in place. Make sure that the part datum feature is firmly in contact with the gage datum element before and after tightening the two Go thread gage elements.

2. Insert the counterbore clearance hole gage. H should be 0.442-in. diam, derived as follows:

$$\begin{array}{r}0.452\text{-in. diam at MMC}\\ \text{minus } \underline{0.010\text{-in. positional tolerance specified at MMC}}\\ 0.442\text{-in. } H \text{ basic diam}\end{array}$$

C should be 0.990-in. diam, derived as follows:

$$\begin{array}{r}1.000\text{-in. diam}\\ \text{minus } \underline{0.010\text{-in. positional tolerance specified at MMC}}\\ 0.990\text{-in. } G \text{ basic diam}\end{array}$$

3. For acceptance, all three gage elements should go into the part at the same time.

Three plug gages are also required: a 0.447-in.-diam Go gage, a 0.452-in.-diam Not-Go gage, and a Not-Go thread gage. The two thread gages that enter bushing B are Go gages.

Less Critical (MMC) Part Datum Features

Figure 4-4A shows a part that has MMC specified for the 0.501–0.502-in.-diam part datum-feature. This datum feature has *also* been allowed a perpendicularity tolerance of 0.001 in. (at MMC) from datum plane A (which must be identified on symmetrical parts).

The functional gage (Fig. 4-4B) for this part may take into account both the MMC size of the part datum feature and the perpendicularity tolerance allowed on that feature by using a 0.500-in.-diam pin D as the fixed datum gage pin. This is determined by subtracting the 0.001-in. perpendicularity tolerance from the MMC size (0.501 in.) of the part feature. The final gage is shown in Fig. 4-4B, and the four fixed-clearance hole gage elements will be:

$$\begin{array}{r}0.380\text{-in. holes at MMC}\\ \text{minus } \underline{0.005\text{-in. positional tolerance when holes are at MMC}}\\ 0.375\text{-in. diam } G \text{ gage pins (4 required)}\end{array}$$

For this operation, four plug gages are required: a 0.501-in.-diam Go gage, a 0.502-in.-diam Not-Go gage (for datum B), a 0.380-in.-diam Go gage, and a 0.390-in.-diam Not-Go gage for the clearance holes.

If the part drawing did not specify a 0.001-in. perpendicularity tolerance for the part datum feature B, the gage would contain a 0.501-in.-diam D

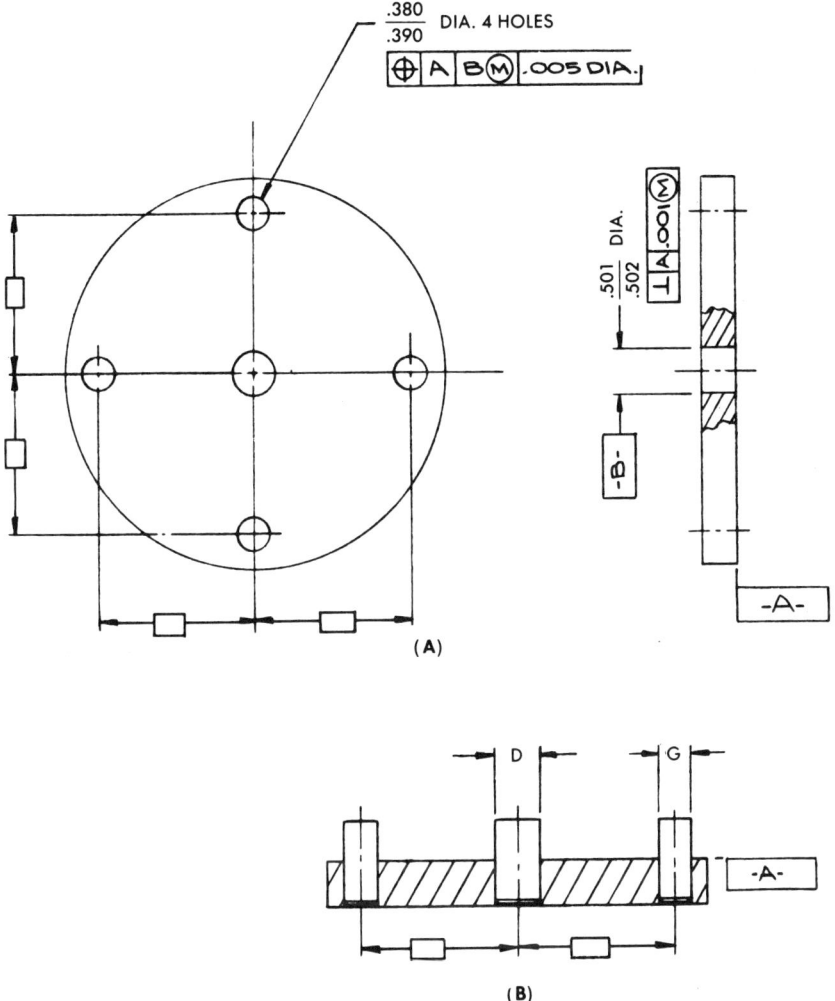

Fig. 4-4. A less critical (MMC) single datum feature and functional gage.

pin which corresponds to the MMC size of the hole. This gage would require perfect perpendicularity when the hole is 0.501 in. (at MMC) and is a more critical MMC callout. This 0.501-in. D gage element would include the 0.501-in. Go gage listed above in this case.

The use of perpendicularity tolerances on part datum features enables the product designer to determine the exact size or the fit allowance of the gage datum element, and thus allows a variety of fits between gage datum elements and part datum-features. MMC gages may "shake" or move on parts when part datum-features are not at MMC and are therefore less critical than RFS gages.

Figure 4-5A shows a part that has MMC specified for the OD part datum feature B. Since no perpendicularity tolerance has been specified for part datum feature B from part datum feature A, the gage diam D will

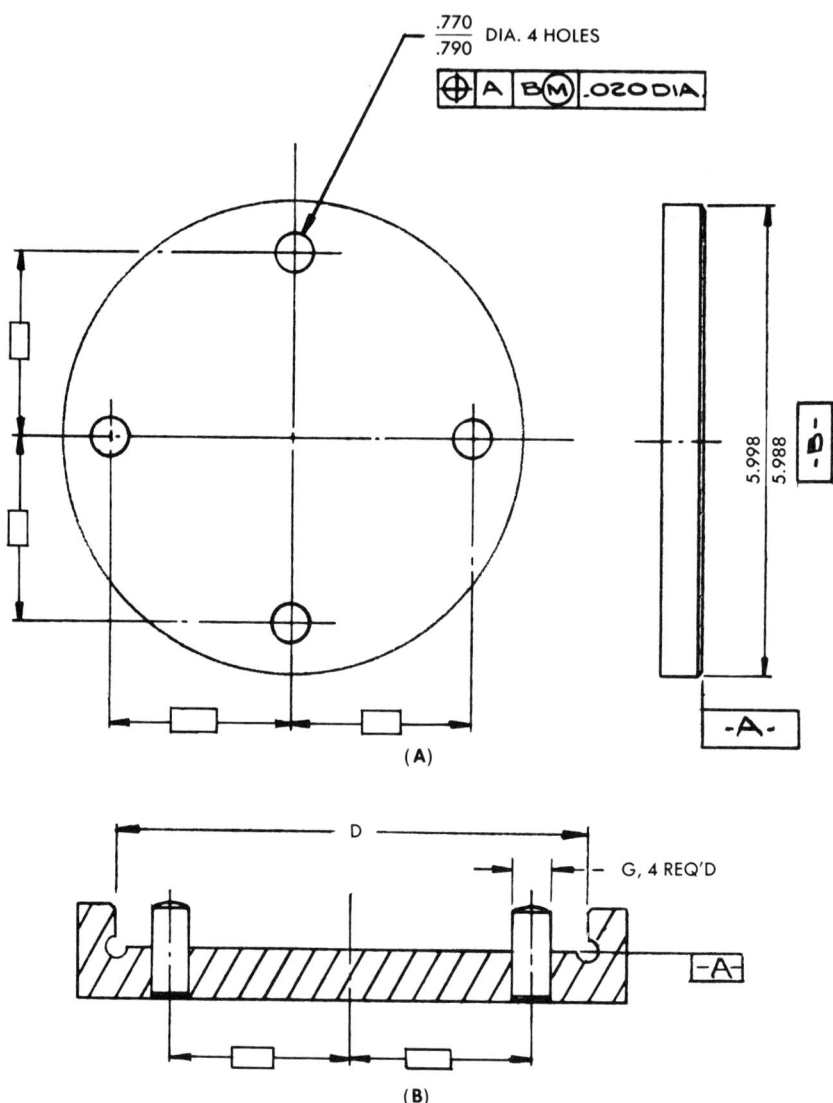

Fig. 4-5. A Less critical (MMC) OD datum feature.

be 5.998 in. which is the MMC size of part diam B (Fig. 4-5A). Gage diam D would be 5.999 in. if part diam B had a 0.001-in. perpendicularity tolerance at MMC specified for part datum feature A. The four fixed gage pins in the gage shown in Fig. 4-5B will be 0.750-in. diam, derived as follows:

$$\begin{array}{l}\text{0.770-in. diam at MMC}\\ \text{minus 0.020-in. positional tolerance specified for the holes}\\ \overline{\text{0.750-in. fixed gage pin size G}}\end{array}$$

For this operation, three gages are required: a 5.988-in.-diam Not-Go ring gage (the 5.998-in. gage datum element is the Go gage); a 0.770-in.-diam Go plug gage; and a 0.790-in.-diam Not-Go plug gage.

Figure 4-6A shows the entire periphery of a non-circular part datum feature modified with the MMC symbol. The gage (Fig. 4-6B) will consist of four rails 4.000 in. × 2.000 in., the MMC size of the part datum feature. Unless otherwise specified on drawings defined per MIL-STD-8, all dissimilar hole patterns on the same part must be gaged as one pattern.

Independent-Hole Patterns

The drawing (Fig. 4-6A) may, however, specify with a note that the "0.390–0.410-in. hole patterns and the 0.878–0.890-in. hole patterns may be gaged separately." Such a note, if added, would result in two gages; one for the 0.390–0.410-in. hole pattern (Fig. 4-7A) and one for the 0.878–0.890-in. hole pattern as shown in Fig. 4-7B. This note would allow some independence of the hole patterns, and the two gages would not be as restrictive as the single gage shown in Fig. 4-6B. It would be applicable if two separate items were assembled in the 0.390–0.410 and the 0.878–0.890-in. holes.

Plug and Snap Gages Required.
1. 3.990-in. Not-Go snap gage for minimum length of "datum" B
2. 1.990-in. Not-Go snap gage for minimum width of "datum" B
   NOTE: Maximum or "Go" dimensions of "datum" B are incorporated in the receiver gages in Fig. 4-7
3. 0.390-in. Go plug gage
4. 0.410-in. Not-Go plug gage
5. 0.878-in. Go plug gage
6. 0.890-in. Not-Go plug gage

Two Critical Datum Features

Figure 4-8A shows two part datum features, diam B and slot C, which control the radial location of the part features. Since part datum features B and C are modified with MMC, the gage in Fig. 4-8B will have a 0.502-in.-diam fixed gage pin B; and since the part slot, "datum" C, also has a perpendicularity of 0.001 in. to datum surface A, the size of gage element

52 Design Principles for Feature Location and Relation Gaging Chap. 4

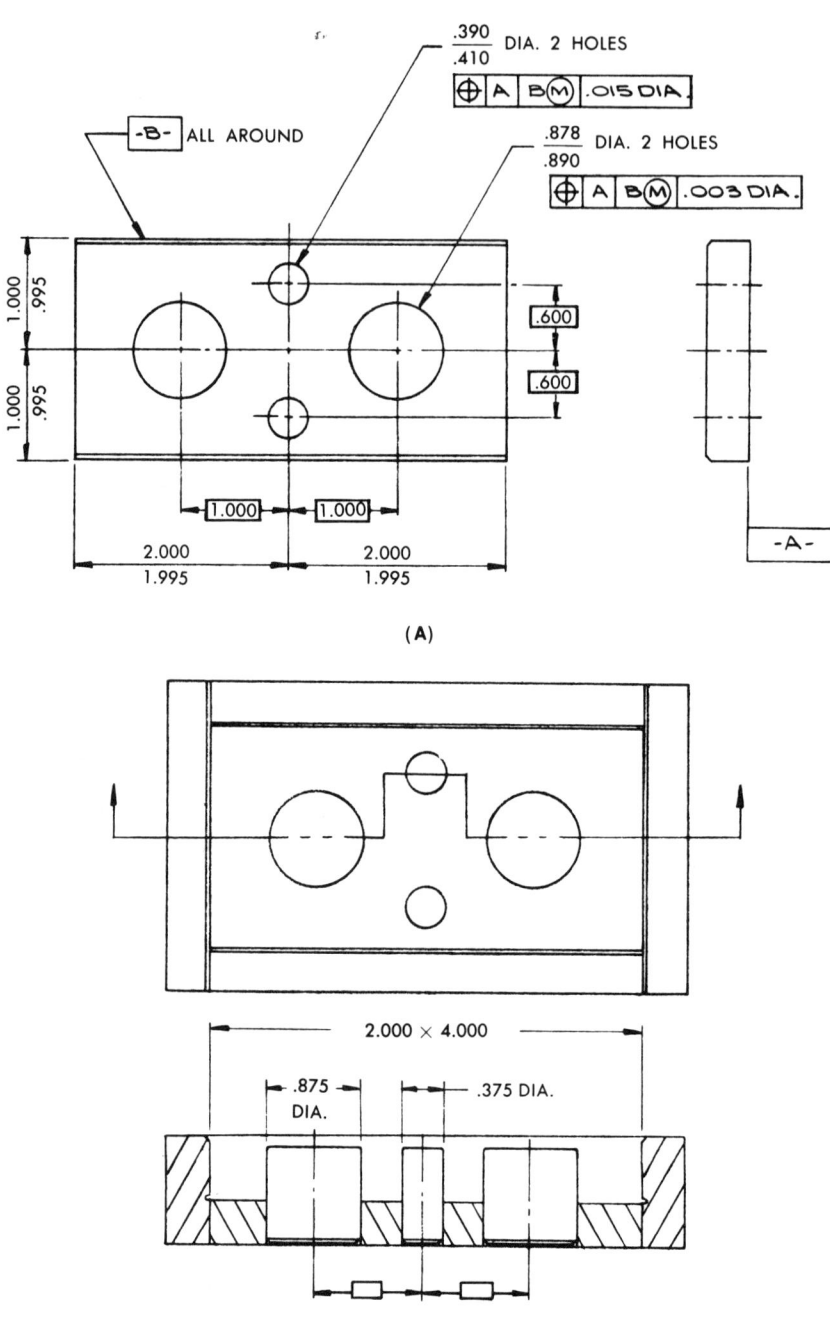

Fig. 4-6. A less critical (MMC) non-circular datum feature.

Chap. 4   Design Principles for Feature Location and Relation Gaging   53

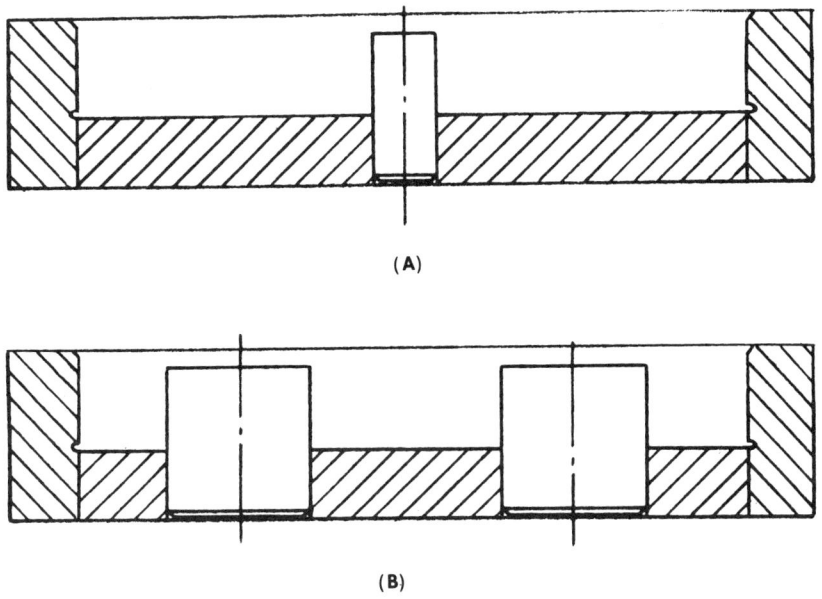

Fig. 4-7. Functional gages for separate gaging of independent part features.

C will be 0.401 in. minus 0.001 or 0.400 inch. The 0.001-in. perpendicularity tolerance is one way of specifying a "shake" gage datum element in that the gage element very likely will never fit too tightly in the part slot. More or less gage "shake" can be easily specified, as previously mentioned, by using various perpendicularity tolerances on part datum features. Thus on the gage, diameter $B = 0.502$ in., and width $C = 0.400$ inch. The three fixed pins G will be 0.375-in. diam determined by subtracting the 0.005-in. positional tolerance at MMC from the MMC hole size of 0.380 inch.

Plug and Ring Gages Required.
For "datum" C:
1. 0.401-in. Go
2. 0.402-in. Not-Go

For "datum" B:
3. 0.504-in. Not-Go

For the three clearance holes:
4. 0.380-in. Go
5. 0.400-in. Not-Go

The OD has a concentricity tolerance of 0.005-in. diam at MMC for part datum feature B at MMC. The ID of the gage datum element that contacts

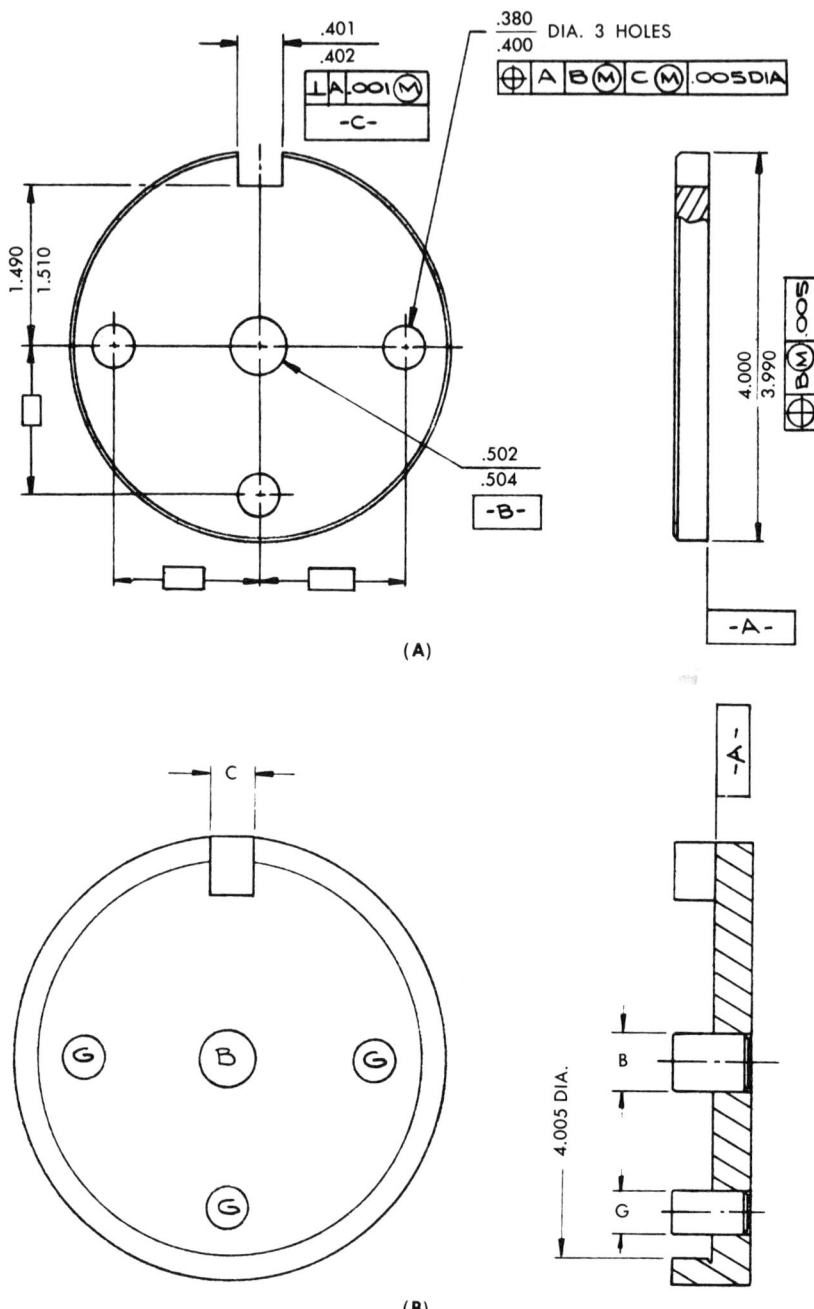

Fig. 4-8. Two critical (MMC) datum features and functional gage.

Chap. 4   Design Principles for Feature Location and Relation Gaging   55

this OD will be

        4.000 in., the MMC size of the OD
plus 0.005 in., concentric tolerance specified at MMC
        $\overline{4.005}$ in., ID of the gage

Two ring gages will be required for the part OD; a 4.000-in. Go gage and a 3.990-in. Not-Go gage.

The gage shown in Fig. 4-8B properly combines several gaging operations on one gage. *If* the positional tolerance specified for the OD, however, were only 0.001 in., and if this OD approached 4.000 in. (its MMC size), and if part datum feature B were 0.504-in., the outside 4.000-in. diam would assume the "datum" function and all holes would be located from this *erroneous "datum" feature*. The gage designer must watch for such pitfalls and use two separate gages when this problem can occur.

Critical Datum-Features, with Independent Hole Pattern

Figure 4-9A shows a hole pattern related to two part datum features, tabs B and B', at MMC.

Design Intent. The part will assemble into a mating configuration so designed that tabs B and B' will be the key alignment features, with surface C the end locator. The six clearance holes will merely hold the part in place at assembly.

Functional Gage Design. Figure 4-9B shows the gage for the part in Fig. 4-9A. The slots in this gage will be 0.4005 in. since part datum features B and B' have a perpendicularity tolerance of 0.005 in. at MMC, derived as follows:

        0.40000 in., MMC size of tabs B, B'
plus 0.00005 in., perpendicularity at MMC specified for B, B^1
        $\overline{0.40005}$ in., gage slot sizes B and B' in Fig. 4-9B

The distance L between gage rails will be 3.210 in., the MMC length of the part. Gage length L is not a gage datum element but is merely a Go length gage. The diam of the six fixed gage pins will be derived as follows:

        0.380 in., MMC size of holes
minus 0.005 in., positional tolerance at MMC
        $\overline{0.375}$ in.-diam gage pins

The part must be placed in the gage against surfaces A and C (Fig. 4-9A). This must be verified by suitable mechanical means.

Plug and Snap Gages Required.
1. 0.4000-in. Go to check "datums" B and B'
2. 0.3980-in. Not-Go to check "datums" B and B'
3. 3.200-in. Not-Go (the 3.210-in. Go gage for dimension is incorporated in the receiver gage in Fig. 4-9A).

56 Design Principles for Feature Location and Relation Gaging Chap. 4

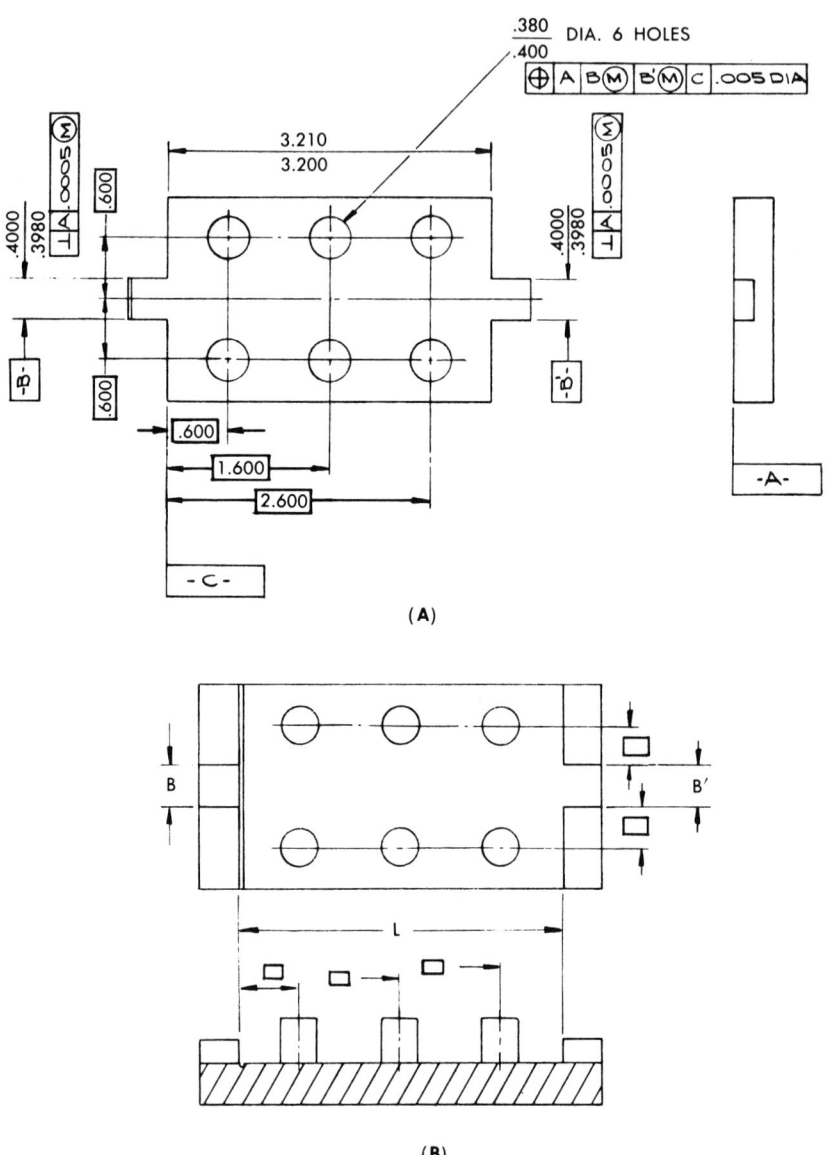

Fig. 4-9. Two critical (MMC) datum features, with independent hole pattern, and functional gage.

Chap. 4 *Design Principles for Feature Location and Relation Gaging*   57

The 0.0005-in. perpendicularity tolerance for "datum" tabs B and B' is incorporated in the functional gage in Fig. 4-9B.

Datum Features Related to Primary Datum Plane

Figure 4-10A shows a part that contains part datum features B and C and primary datum plane A. This method of dimensioning parts from two holes is quite common because it is practical to locate parts on tools that contain dowel pins.

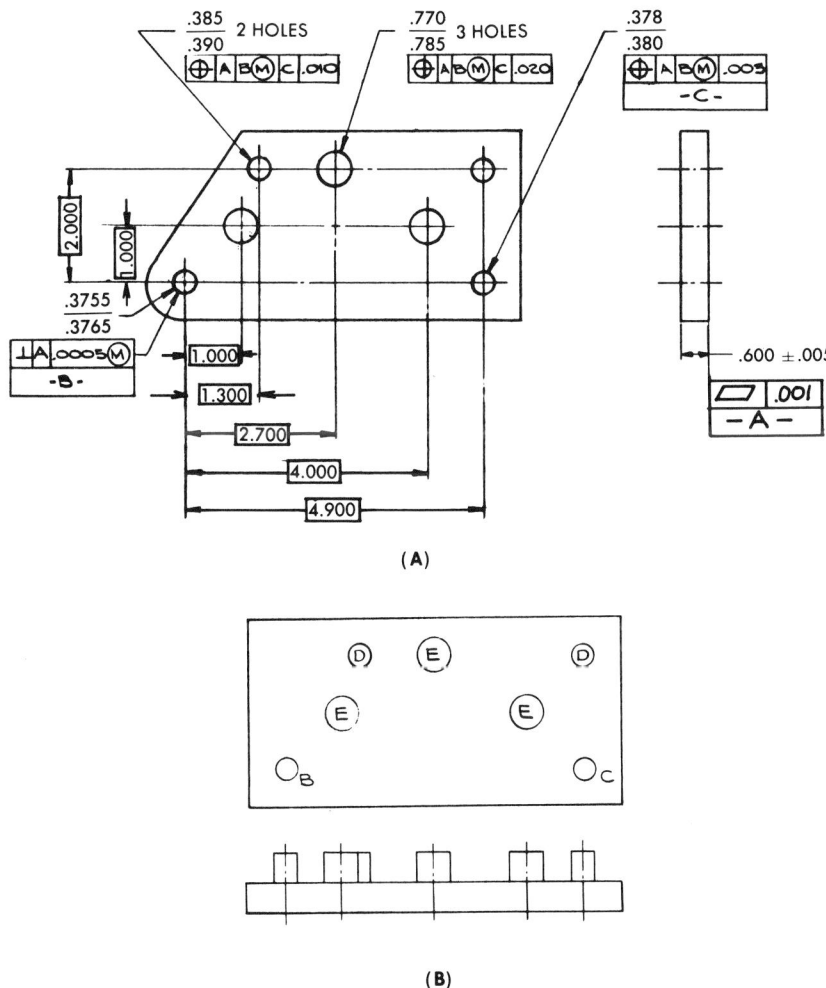

Fig. 4-10. Part with two holes as datum features related to a primary datum, and multi-pin functional gage.

Figure 4-10B shows the multi-pin gage that will meet the drawing requirements. The gage pins are determined as follows:
1. B gage datum element

$$\begin{array}{r} 0.3755\text{-in. diam at MMC} \\ \text{minus } \underline{0.0005\text{-in. perpendicularity tolerance at MMC}} \\ 0.3750\text{-in. diam at } B \end{array}$$

2. C gage datum element

$$\begin{array}{r} 0.378\text{-in. diam at MMC} \\ \text{minus } \underline{0.003\text{-in. positional tolerance at MMC}} \\ 0.375\text{-in. diam at } C \end{array}$$

3. D gage elements

$$\begin{array}{r} 0.385\text{-in. diam at MMC} \\ \text{minus } \underline{0.010\text{-in. positional tolerance at MMC}} \\ 0.375\text{-in. diam at } D \end{array}$$

4. E gage elements

$$\begin{array}{r} 0.770\text{-in. diam at MMC} \\ \text{minus } \underline{0.020\text{-in. positional tolerance at MMC}} \\ 0.750\text{-in. diam at } E \end{array}$$

Plug Gages Required.
1. 0.3755-in. Go   )
2. 0.3765-in. Not-Go) for datum-feature B
3. 0.378-in. Go   )
4. 0.380-in. Not-Go) for datum-feature C
5. 0.385-in. Go   )
6. 0.390-in. Not-Go) for 0.385–0.390-in.-diam clearance holes
7. 0.770-in. Go   )
8. 0.785-in. Not-Go) for 0.770–0.785-in.-diam clearance holes

Three-Hole Pattern and External Datum Feature

Figure 4-11A shows a radial three-hole pattern dimensioned from primary part datum feature A and diam B at MMC.

Design Intent. The part must fit into a sleeve against surface A and be pinned in location through three holes. Part datum feature B must be perpendicular to part datum feature A at MMC.

Functional Gage Design. Several dial indicators could be so mastered that they would all read zero when the part is fully inserted into the gage (Fig. 4-11B). The 2.000-in. ID in the gage is a Go gage that checks both the maximum size of the part OD and its perpendicularity to datum surface A. The three G gage elements will be 0.313-in. diam, determined as follows:

Chap. 4 Design Principles for Feature Location and Relation Gaging   59

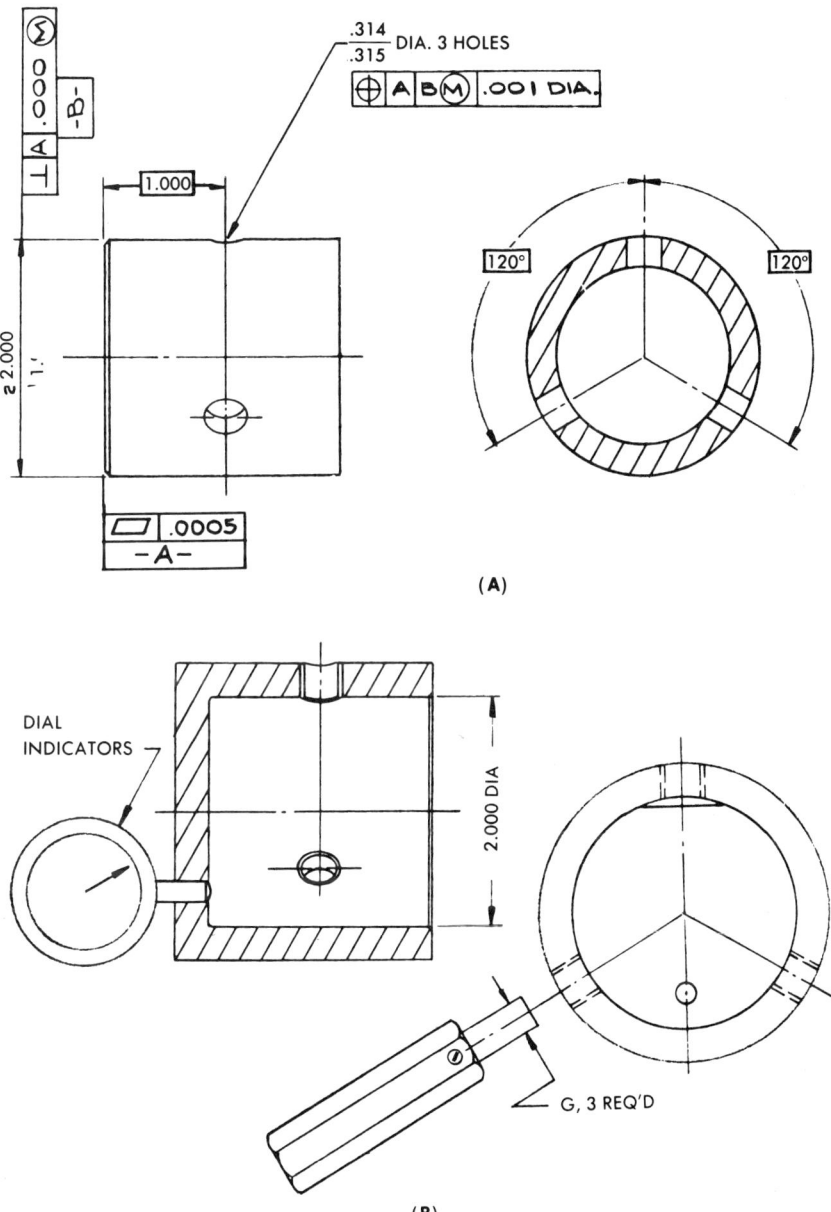

Fig. 4-11. Part with radial three-hole pattern and external datum feature, and functional gage.

0.314-in. clearance hole diam at MMC
minus 0.001-in. positional tolerance specified at MMC
0.313-in. *G* gage basic diam

Plug and Ring Gages Required.
1. 0.314-in. Go
2. 0.315-in. Not-Go
3. 1.995-in. Not-Go snap gage

The same gage design technique is used regardless of the number of radial holes in the pattern. The same basic gage would also be used if the three holes shown in Fig. 4-11*A* were inclined at some basic angle, other than 90 deg to part datum feature *A*, except that the gage bushings would be placed at this same angle.

Three-Hole Pattern and Internal Datum Feature

Figure 4-12*A* shows a radial three-hole pattern dimensioned from part datum feature *A* and ID *B* at MMC.

Design Intent. The part must fit over a plug against surface *A* and be fastened with three bolts. Part datum feature diam *B* need not be perfectly square at MMC since it is allowed 0.001-in. perpendicularity at MMC to part datum feature *A*.

Functional Gage Design. The gage (Fig. 4-12*B*) is a shouldered plug with an OD of 1.549 in., determined as follows:

1.550 in., the MMC size of the part datum feature
minus 0.001 in., perpendicularity tolerance specified at MMC
1.549 in., gage element diam *B*

For acceptance, part datum feature *A* must fully contact the gage datum element *A*. Suitable mechanical verification is required. The *G* gages will be 0.316-in. diam. The same basic gage would also be used if the three holes shown in Fig. 4-12*A* were inclined at some basic angle other than 90 deg to part datum feature *A*, except that the gage bushings would be placed at this same angle.

Plug Gages Required.
1. 1.550-in. Go
2. 1.552-in. Not-Go
3. 0.320-in. Go
4. 0.330-in. Not-Go

Cylindrical Part with Two-Pin Patterns.

Figure 4-13*A* shows a cylindrical part with one two-pin pattern and a four-hole pattern. The part datum feature for these patterns is the axis of the 1.000–1.005-in. ID at MMC. The two end surfaces have not been designated as part datum features and will only be contacted at one high point when the gage (Fig. 4-13*B*) is placed on the part.

Chap. 4 Design Principles for Feature Location and Relation Gaging

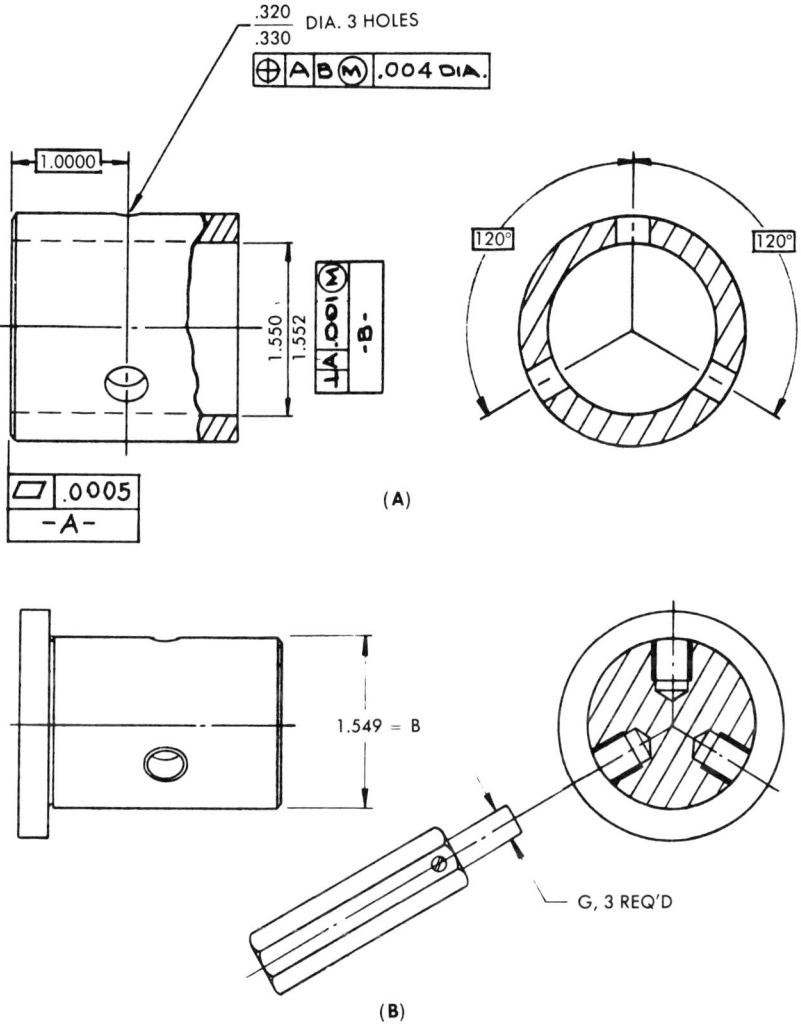

Fig. 4-12. Part with radial holes and an internal datum feature, and functional gage.

Design Intent. The mating part passes through part datum feature diam A (Fig. 4-13A). The two patterns must be located from the ID of part datum feature A when this feature is at MMC (1.000 in.), and the part datum feature must be straight at MMC (when it is finished to 1.000-in.). The drawing note states that the two-pin and four-hole patterns are not radially related (they do not have to align for the parts to function at assembly).

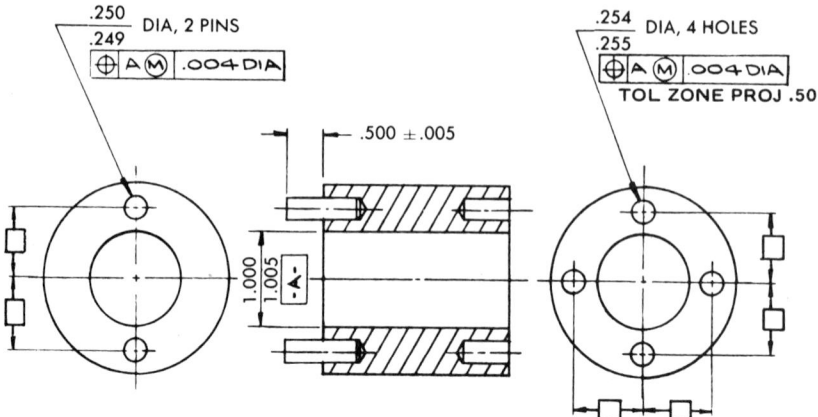

NOTE: RADIAL ALIGNMENT OF TWO FEATURE PATTERNS NOT CRITICAL

(A)

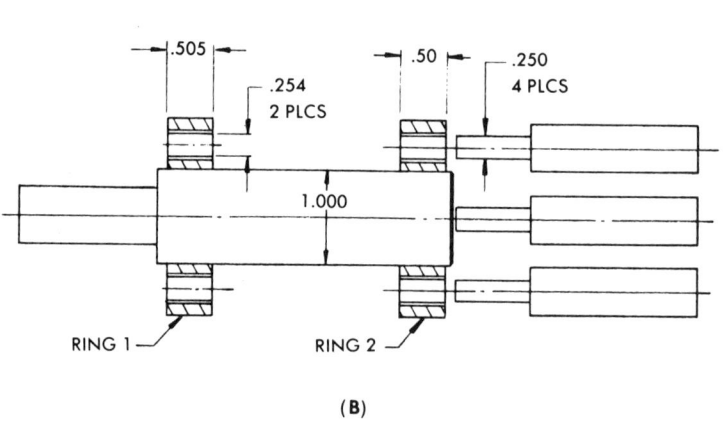

(B)

Fig. 4-13. A part with two feature patterns and functional gage.

Functional Gage Design. The gage datum element diam (1.000 in.) is the MMC size of part datum feature A. Two 0.254-in. ID bushings are placed on the gage ring *1* that fits closely over the 1.000-in. receiver diam. Four 0.250-in.-diam gages will fit through the four close-fitting bushings on the gage ring *2* element that also fits closely over the 1.000-in. receiver gage datum element diam. The two rings, which control the two- and four-hole patterns on the part, are free to rotate on the 1.000-in. receiver diam since pattern alignment is not critical. Ring *1* is at least 0.505-in. thick, this being the maximum height of the two 0.250–0.248-in.-diam

Chap. 4 *Design Principles for Feature Location and Relation Gaging*   63

pins, and ring *2* is at least 0.50 in. thick, the maximum thickness of the mating part. If the drawing did not state that the two-feature patterns could rotate (or misalign), the two rings should be aligned or oriented with keyways or pins.

Plug and Ring Gages Required.
1. 0.250-in. Go ring
2. 0.249-in. Not-Go ring
3. 1.005-in. Not-Go plug (The 1.000-in.-diam receiver gage element checks the 1.000-in. "go" requirement).
4. 0.254-in. Go Plug
5. 0.255-in. Not-Go plug

Two Datum Features with Two Feature Patterns

Figure 4-14*A* shows a part similar to the one shown in Fig. 4-13*A*, except for having the two-pin and four-hole patterns perpendicular to primary part datum features A and A'. The ID portions marked with X and XX are the useful portions of part datum feature C. X and XX represent the maximum depth of entry of the mating parts at assembly and would be toleranced dimensions on the part drawing.

Design Intent. The mating part does not pass through part datum-feature C; instead, two mating parts of lengths X and XX assemble to each end of the part. The two-pin pattern must be located from the small X depth of part datum feature C when it is at MMC. The four-hole pattern must be located from the XX depth of part datum feature C' when it is at MMC. The two-pin and four-hole patterns are not radially related, so two separate gages may be used.

Functional Gage Design. Two separate gages are required: one for the two-pin pattern shown in Fig. 4-14*B*, and one for the four-hole pattern shown in Fig. 4-14*C*.

Plug Gages Required. The plug gages are the same as those used to inspect the part shown in Fig. 4-13*A*. If the drawing in Fig. 4-14*A* did not state that the two-feature patterns on the part were free to rotate ("rotational alignment is not critical"), the two gages of Figs. 4-14*B* and 4-14*C* would need to be combined and held in strict alignment when used. If the drawing stated that the two patterns must be aligned within ± 1 deg, then the two gages would have to be combined so that one could rotate in relation to the other out of the perfect alignment shown on the drawing by only ± 1 deg.

Two Radial Patterns of Pins and Slots

Figure 4-15*A* shows two radial patterns of features (pins and slots).

Design Intent. The radial orientation of these part features is critical in relation to part datum feature B, but their location from part datum

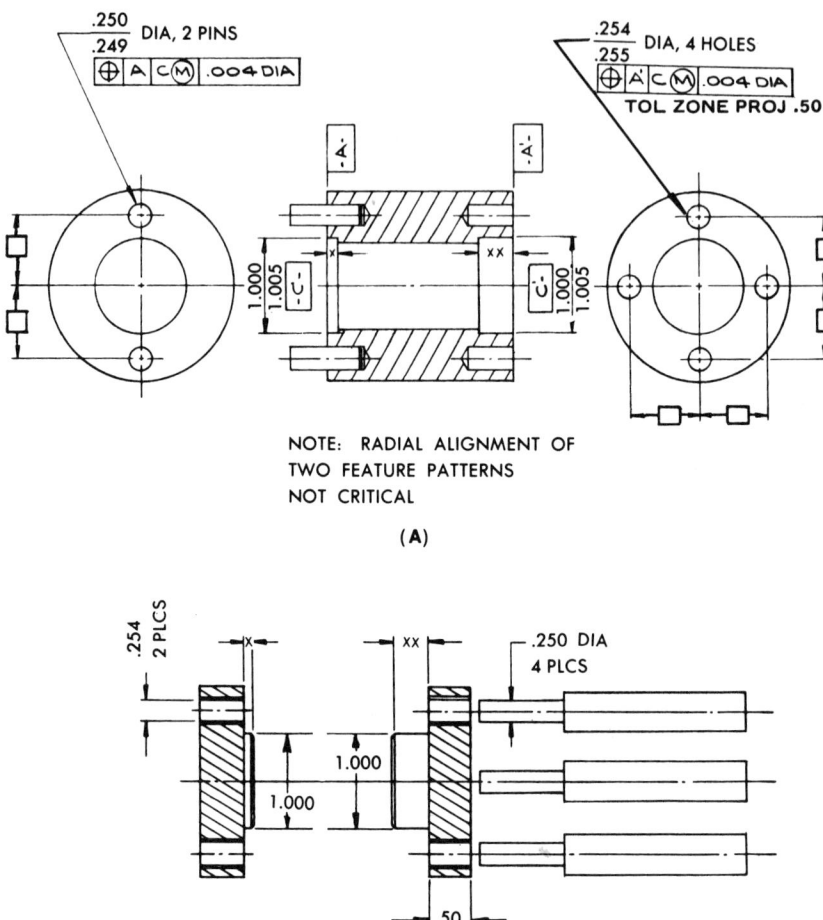

Fig. 4-14. Part with two datum features and two feature patterns, with functional gages.

feature *A* is not critical. The requirement has been presented in note form on the drawing because there is no ANS Y14.5 symbol for a positional tolerance in *width*. The OD of the part must be positioned within 0.005-in. diam at MMC to part datum feature *B* at MMC.

Functional Gage Design. The gage (Fig. 4-15*B*) is one unit since the pins and slots are related and are actually one pattern of features dimensioned from part datum feature *B* at MMC.

1. The gage datum-element *B* is 1.9995 in. because the part datum feature

Chap. 4 Design Principles for Feature Location and Relation Gaging   65

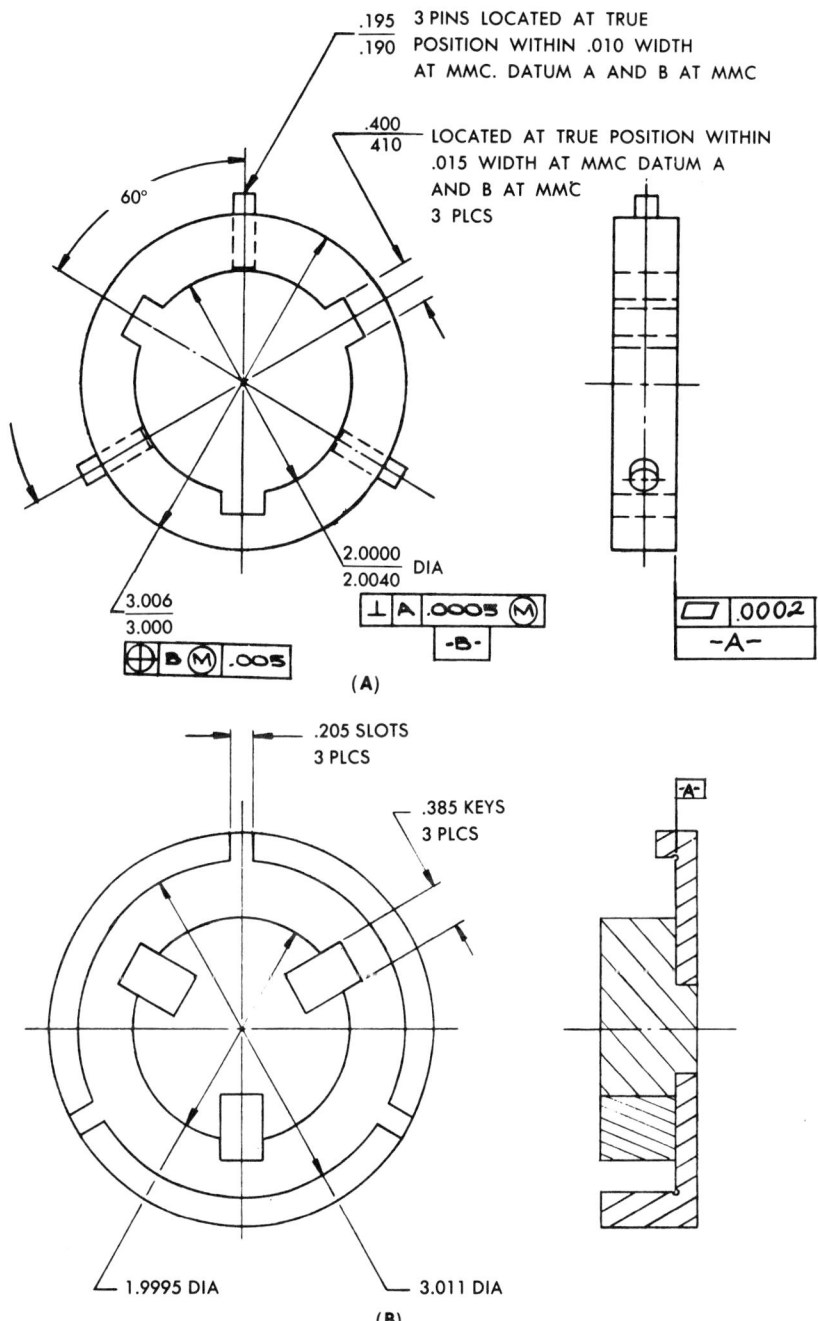

Fig. 4-15. Part with radial pins and slots, and functional gage.

has a perpendicularity tolerance of 0.0005 in. in relation to part datum feature A at MMC.

 2.0000 in., part datum feature B at MMC
minus 0.0005 in., perpendicularity tolerance at MMC
 $\overline{1.9995}$ in., gage datum element

2. Pin location (radial) is checked with three 0.205-in.-wide slots.

 0.195 in., MMC of pin
plus 0.010 in., positional tolerance at MMC
 $\overline{0.205}$ in., gage element slot

3. Slot locations are checked with three 0.385-in. contacts.

 0.400 in., MMC of slots
minus 0.015 in., positional tolerance of slots
 $\overline{0.385}$ in., gage key element

4. Part OD positional tolerance is checked with the 3.011-in.-diam gage ID, determined as follows:

 3.006 in., MMC of part OD
plus 0.005 in., positional tolerance allowed at MMC
 $\overline{3.011}$ in., gage element diam

Plug Gages Required.
1. 0.195-in. Go
2. 0.190-in. Not-Go
3. 0.400-in. Go
4. 0.410-in. Not-Go
5. 2.0000-in. Go (required because functional gage is 1.9995 in.)
6. 2.0040-in. Not-Go
7. 3.006-in. Go (required because functional gage is 3.011-in.)
8. 3.000-in. Not-Go

Review of Basic Principles
1. Add specified positional tolerance to MMC size of male part features to obtain basic female size of the functional gage element.
2. Subtract positional tolerance specified for MMC size of female part features to obtain basic male size of the functional gage element.
3. Part datum features modified with RFS must be "centered" in the gage.
4. Part datum features modified with MMC must be contacted by gage datum elements that are the MMC size of these part datum features.
5. Part datum features modified with MMC that have perpendicularity tolerances also modified with MMC, are contacted by gage datum elements that never freeze on the part datum feature. Such a gage could always shake on the part.

Chap. 4   Design Principles for Feature Location and Relation Gaging   67

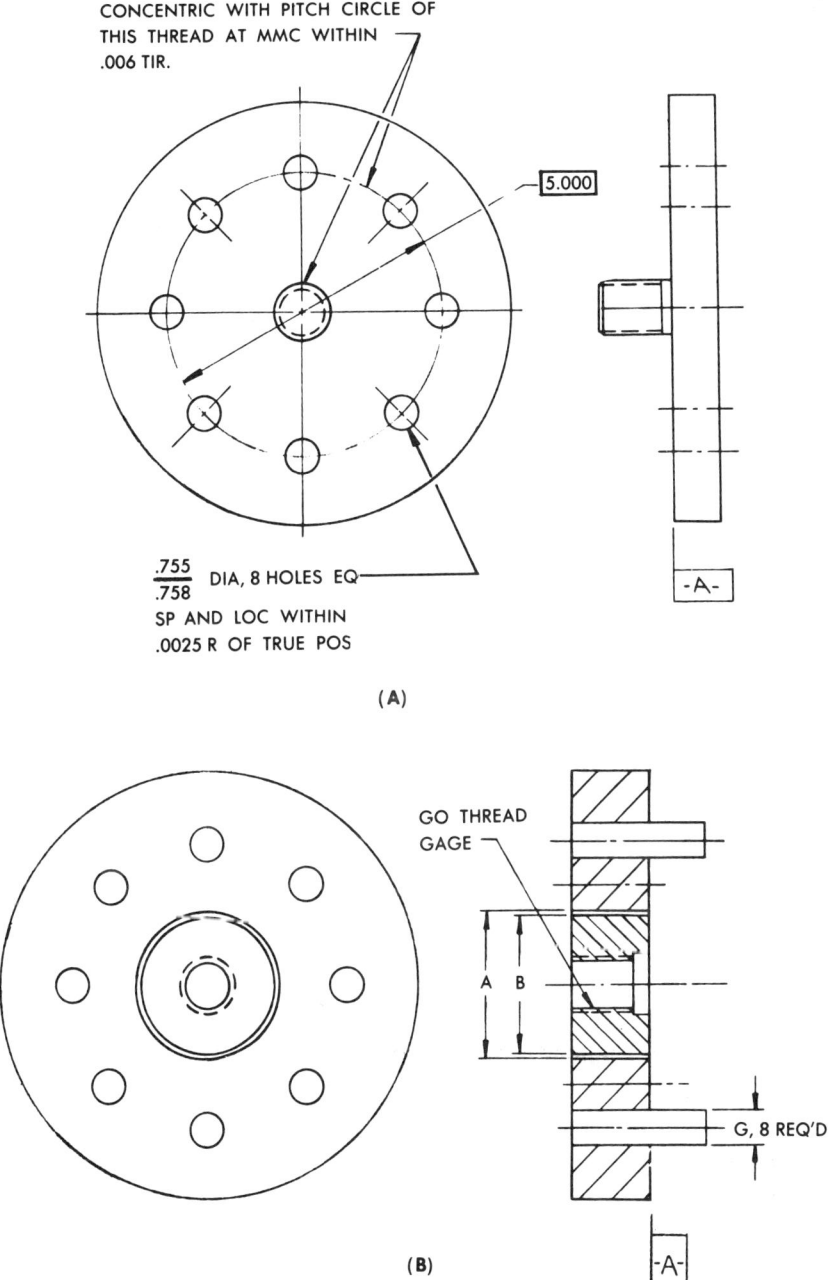

Fig. 4-16.   A non-critical part datum feature and functional gage.

Non-Critical Part Datum Features

Figure 4-16*A* shows a hole pattern on a 5.000-in. basic bolt circle. The concentricity tolerance of 0.006-in. TIR allows a specific independence of 0.006-in. diam between the hole pattern and the part datum feature or stud. The eight-hole pattern must be gaged as one pattern of features with a feature relation gage, but this gage should be separated from the gage element that contacts the part datum feature or stud so that the hole pattern is allowed a 0.006-in. independence from this "datum" stud.

Figure 4-16*B* shows the gage. The difference between *A* and *B* diameters on the two gage elements is 0.006 in. and this allows the hole pattern a 0.006-in.-TIR (or independence) from the center stud. Each of the fixed gage pin elements will be 0.750 in. diam, determined as follows:

$$0.755 \text{ in., specified MMC size of hole}$$
$$\text{minus } \underline{0.005 \text{ in., } (2 \times 0.0025\text{-in. radius specified at MMC})}$$
$$0.750 \text{ in., fixed gage pin size } G$$

Special Cases

Figure 4-17*A* shows a five-hole pattern with the center hole as "datum" for the four-hole pattern. This five-hole pattern, however, is not precisely related to the outside of the part. To indicate this relaxed relationship the five-hole pattern is located, somewhat loosely, with the 0.170-in. minimum dimension. This method of dimensioning indicates that a feature relation gage is required that does not contact the part periphery.

Figure 4-17*B* and *C* shows the two gages required for this part. The larger center fixed gage pin element will be 1.000 in. diam (1.001 in. less the 0.001-in. perpendicularity tolerance on this hole from datum plane *A*). The four-hole pattern of fixed gage pins in Fig. 4-17*B* will be 0.190 in. diam:

$$0.200 \text{ in., MMC hole size}$$
$$\text{minus } \underline{0.010 \text{ in., positional tolerance at MMC}}$$
$$0.190 \text{ in., fixed gage pin size } G$$

The pattern is checked for location on the part with a Not-Go gaging element shown in Fig. 4-17*C* which must not go into each of the four corner holes and over the periphery of the part. The distance between pins on this gage is the 0.170-in. minimum distance specified on the drawing in Fig. 4-17*A*.

Plug Gages Required
1. 0.200-in. Go
2. 0.210-in. Not-Go
3. 1.001-in. Go
4. 1.005-in. Not-Go

Datums Not Modified with RFS or MMC Symbols. Figure 4-18*A* shows a pattern of holes dimensioned from datum features *A*, *B*, and *C*

Chap. 4 Design Principles for Feature Location and Relation Gaging

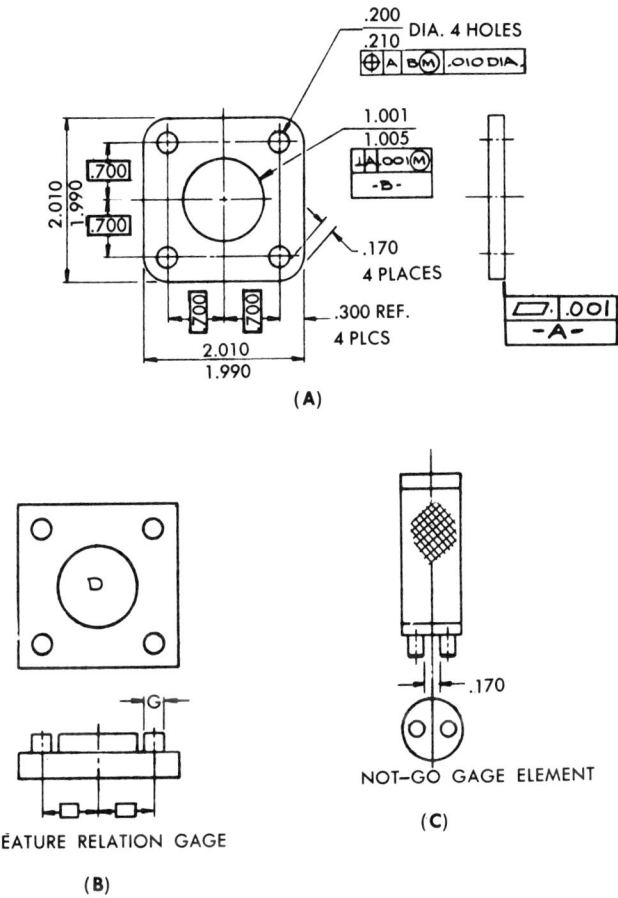

Fig. 4-17. Non-critical pattern location and functional gages.

with no RFS or MMC modifying symbols. The "datum corner" should be identified if the part is symmetrical.

Figure 4-18B shows the feature pattern location and relation gage. To be accepted, the part must go on the six 0.500-in.-diam gage pin elements shown, contact the three tangent gage datum locators, and rest firmly against surface A on the gage base. Suitable mechanical verification must be utilized to assure this. Go and Non-Go plug gages and Go and Not-Go snap gages would, of course, be required for the hole diams and part size.

Phantom Gages

The phantom gage technique specifies an acceptable part by defining the acceptance gage and part on one drawing. This approach demands a

70 Design Principles for Feature Location and Relation Gaging Chap. 4

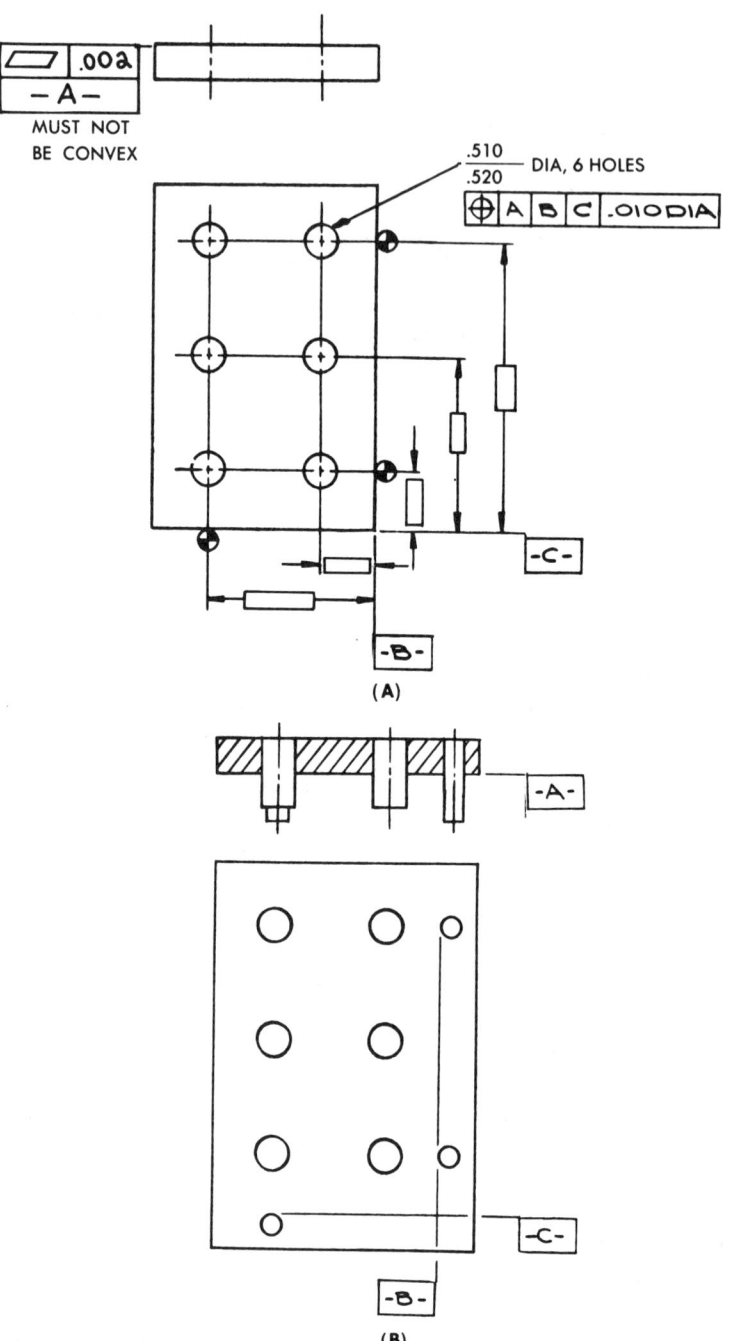

Fig. 4-18. Dimensioning from three part datum surfaces, and functional gage.

Chap. 4 Design Principles for Feature Location and Relation Gaging   71

single interpretation and is particularly effective when used to define extremely small parts suitable for optical chart gage inspection and any part inspected with a functional-type gage.

The proper use of this technique demands that the designer determine and specify one exact acceptance criterion for his part, and this necessitates his knowledge of gage and optical chart design. Fortunately, this is not difficult because functional gages resemble mating parts at their most critical interchangeability (interface) size. The designer can gain this knowledge by viewing and determining mating part relationships during the design layout stage of development. Similarly, optical chart gages simply define maximum and minimum part size and feature relationships, again obtainable from the design layout. The phantom gage is not a universal dimensioning and tolerancing panacea, but it is a direct, uncomplicated, pictorial approach to both product and gage definition.

Functional Gaging. Figure 4-19*A* shows a part entirely defined with phantom-line tolerance zones and Fig. 4-19*B* shows the corresponding receiver gage. The following is an explanation of this new technique.

1. Draw the part to scale, and dimension as follows (see Fig. 4-19*A*):
    a. Use basic dimensions to *locate* features (holes, bosses, etc.).
    b. Use maximum size dimensions to define internal features.
    c. Use minimum size dimensions to define external features.
2. Superimpose one phantom tolerance zone to outline the maximum size of the external features and the minimum size of the internal features. Such an outline, determined from the design layout, should represent the mating part (not shown) at its most critical interchangeability (interface) size.
3. Size and relate these phantom lines to the maximum and minimum part configuration.
4. The phantom lines *are* the basic outline of the Go functional gage. See Fig. 4-19*B*.
5. One final step is required. The designer will specify maximum and minimum size values on the part features that require subsequent and independent Not-Go gaging operations. The 1.000-in. minimum and 2.050-in. maximum diams on Fig. 4-19*A* specify this type of information.

Optical Chart Gaging. Small, relatively thin parts are usually quite suitable for optical inspection since they may be magnified numerous times. Most optical projectors have available magnifications of 10, 20, 31.25, 50, 62.50, or 100. This phantom gaging technique is also particularly applicable to parts entirely defined with phantom line contour tolerance zones because these zones describe the exact optical projector chart gage outlines.

The following description illustrates the use of this technique:

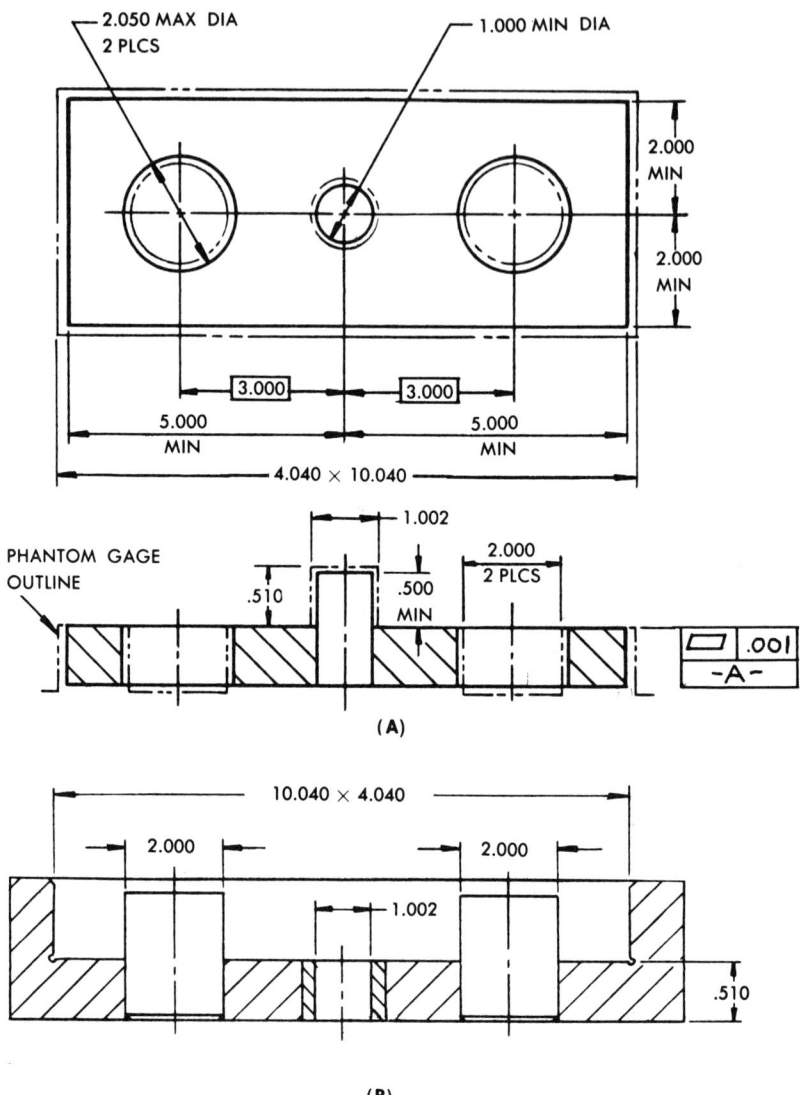

Fig. 4-19. Part entirely defined with phantom-line tolerance zones, and functional gage.

1. Draw the part to scale, using basic, nominal dimensions. The method of dimensioning is not critical, but all features must be nominally located and sized.
2. Superimpose two tolerance zones, designated by phantom lines, on either side of the basic part contour. These zones define the limits of

Chap. 4   Design Principles for Feature Location and Relation Gaging   73

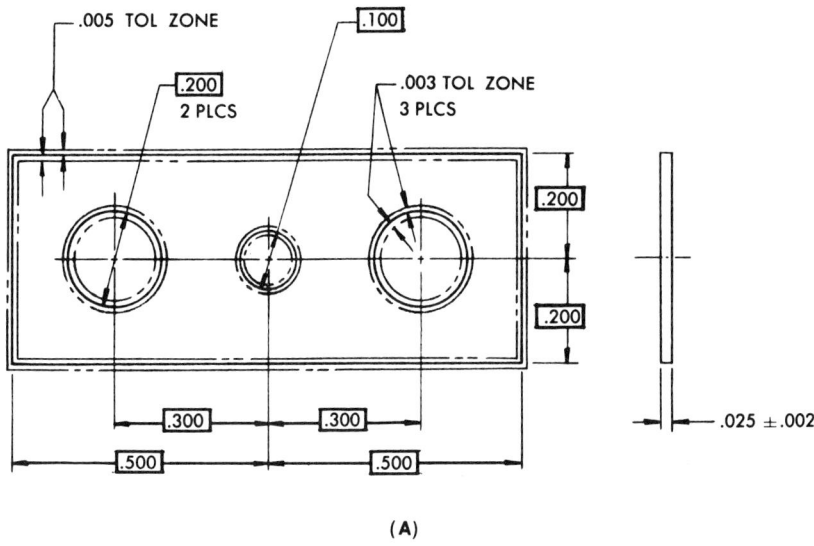

(A)

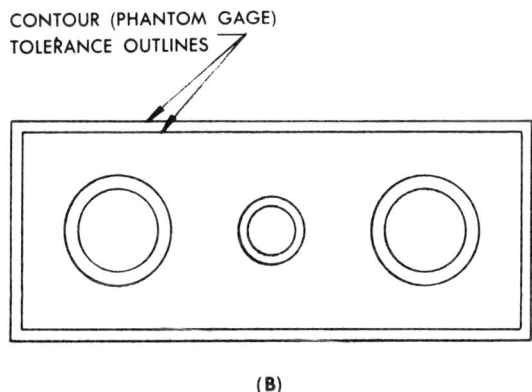

(B)

Fig. 4-20. Phantom gage dimensioning and tolerancing, with chart gage.

size and the location of the part features and need not be to scale. See Fig. 4-20A.
3. With specific dimensions, size and relate these phantom lines to the basic part configuration.
4. The phantom lines are the basic chart gage that will be placed on the viewing screen of an optical projector. The final part shadow, when magnified to the same scale as the chart gage, must lie within the chart gage tolerance zones. Fig. 4-20B illustrates the single chart gage design demanded by the part defined in Fig. 4-20A.

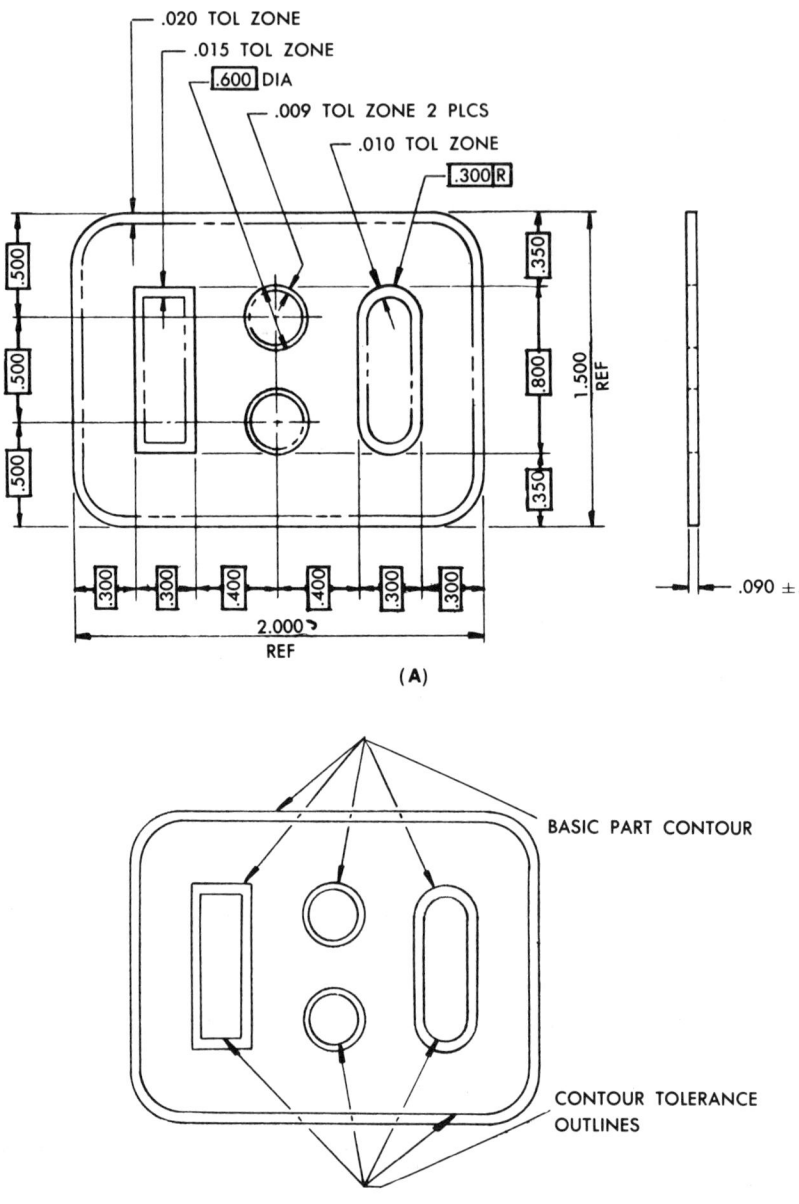

Fig. 4-21. Phantom gage dimensioning and tolerancing (single tolerance zone), with chart gage.

Figure 4-21A shows a part contour defined with basic dimensions and only one phantom tolerance zone. In this case, the chart gage is made by using the basic contour and the phantom tolerance zone as the lines to be reproduced on the chart gage. Figure 4-21B illustrates the final chart gage required to inspect the part shown in Fig. 4-20A.

Figure 4-22A shows two diameters positionally toleranced to a third datum feature axis.

Design intent requires that the 0.600–0.602-in. and the 1.000–1.002-in. part feature diameters be simultaneously related to part datum feature A. The functional gage (Fig. 4-22B) is determined as follows:

D 1.041 in., the MMC size of part datum-feature A
G 0.605 in. (0.602 in. + 0.003 in.)
G' 1.004 in. (1.002 in. + 0.002 in.)

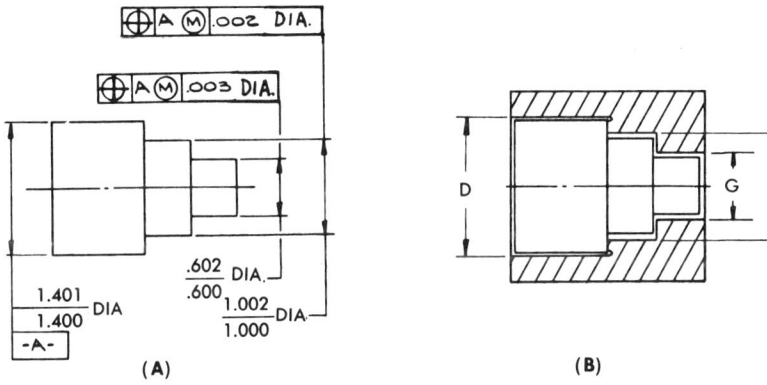

Fig. 4-22. Part and gage for tolerance.

Plug gages are required as follows:
1.400-in. Not-Go
0.602-in. Go
0.600-in. Not-Go
1.002-in. Go
1.000-in. Not-Go

Figures 4-23A and B show the two separate gages required if the drawing in Fig. 4-22A had stated that the 0.602–0.600 in. and 1.002–1.000-in.-diams should be gaged separately. Gage elements D, G, and G' will not change in size.

Figure 4-24A shows a part containing two diams, each of which must be concentric at MMC to datum A at MMC. The note, "The two diameters shall be gaged separately," indicates that two independent gaging opera-

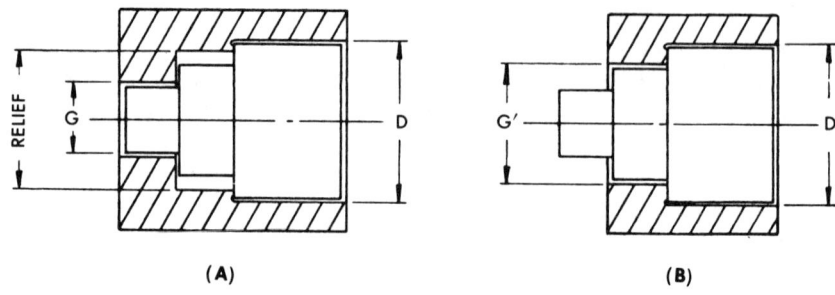

Fig. 4-23. Gages and gaging operation for separately gaged positional tolerances.

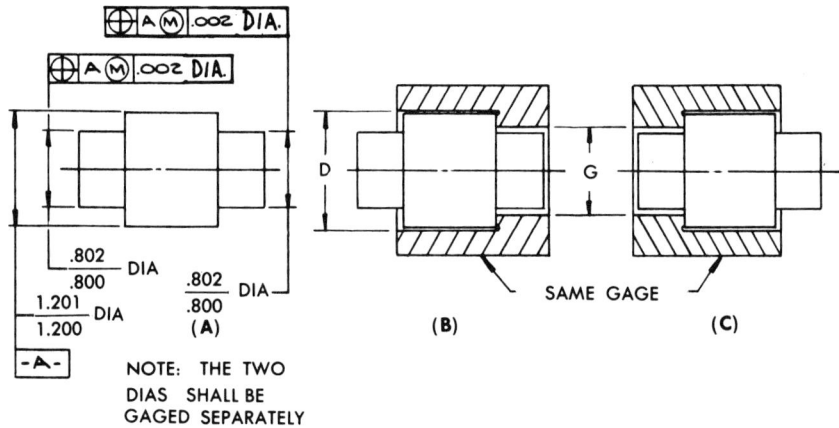

Fig. 4-24. Two independent gaging operations with the same gage.

tions (each using the same gage in this case) will be required, each using part datum feature A at MMC. Figure 4-24B shows the gage that checks both diameters. The same gage is shown in Fig. 4-24C checking the other diameter. Diameter D on the gage will be 1.021 in., the MMC size of part datum A. Diameter G on the gage will be 0.804 in., derived as follows:

0.802 in., part diam at MMC
plus 0.002 in., tolerance at MMC
0.804 in., G gage element size

Plug gages are required as follows: (1) a 1.200-in. Not-Go gage (the 1.201-in. Go is on the functional gage); (2) an 0.802-in. Go gage; and (3) an 0.800-in. Not-Go. If the note requiring separate gage operations were not on the drawing, a split two-piece gage would have to fit over all three part diams at once.

Chap. 4 Design Principles for Feature Location and Relation Gaging 77

Figure 4-25A shows symmetry is a positional tolerance. The symmetry requirement could be called "positional tolerance 0.010 in. wide" and be a valid specification. Symmetry may be used when the toleranced part feature is located on the same axis as the part datum-feature. Slot squareness and parallelism to part datum-features *A* and *B* are all included in the 0.010-in.-wide tolerance zone.

The gage shown in Fig. 4-25B indicates the approximate form of the mating part. The median plane of the slot must lie between two planes, perpendicular to part datum-feature *A*, 0.010 in. apart at MMC, and equi-distant from the median plane of the part datum-feature *B* when this datum-feature is at MMC. The gage datum element *D* (Fig. 4-25B) will be 1.400 in., the MMC size of part datum-feature *B*. Fixed-size gage

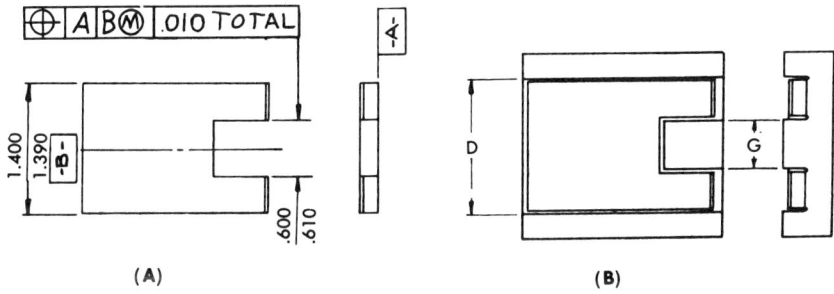

Fig. 4-25. Part definition and symmetry gage.

element *G* that checks the 0.010-in. symmetry tolerance at MMC is derived as follows:

0.600 in., the MMC size of the part feature
minus 0.010 in., the symmetry tolerance specified at MMC
0.590-in.-wide gage element *G* "centered" on the same
median plane as gage element *D*.

Plug gages are required as follows. For the part datum-feature, a Go plug gage is not required because the 1.400-in. gage element *D* checks the MMC size of the part datum-feature. Thus, only a 1.390-in. Not-Go snap gage is required. For the symmetrical part feature, a 0.600-in. Go gage and a 0.610-in. Not-Go gage are required. Had a symmetry tolerance of 0.000-in. width at MMC been specified, the *D* gage element would be the same (1.400 in.), but the *G* gage element would become 0.600 in., or the MMC size of the part feature. This *G* element would replace the 0.600-in. Go plug gage and would require perfect symmetry at MMC. It would also allow a symmetry tolerance only when the part slot was wider than 0.600 inch.

CHAPTER 5

GAGING FORM TOLERANCES

The following form tolerances cannot be conveniently gaged with the fixed-element functional gages discussed in this text:
1. Flatness
2. Straightness (of surfaces)*
3. Angularity (of surfaces)*
4. Perpendicularity (of surfaces)*
5. Parallelism (of surfaces)*
6. Roundness*
7. Profile Tolerances*
8. Runout
9. Cylindricity

Form tolerances modified with the RFS specification cannot be gaged with fixed-element functional gages. Only form tolerances modified with MMC will be considered in this chapter.

Figure 5-1*A* shows a straightness tolerance of 0.002 in. required at MMC. Design intent requires that the pin may be 0.002 in. out-of-straight when it is 1.000-in. diam (or at MMC). This allows the unrestrained effective diam of the pin to be 1.002 in. (Table 5-1).

The functional gage is a sleeve-type with a 1.002-in.-diam ID (1.000-in.

---
*Can be checked with optical chart gages if the parts are of the proper size and tolerance.

Chap. 5 Gaging Form Tolerances

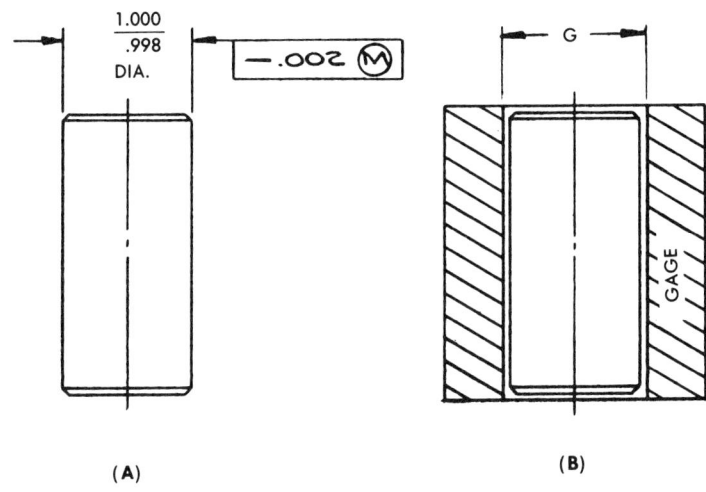

Fig. 5-1. Part and gage for straightness tolerance.

plus 0.002-in. tolerance) at G. Two snap gages are required: a 1.000-in. Go gage, and a 0.998-in. Not-Go gage.

Table 5-1. Relation Between Pin Diam and Straightness Tolerance

Finished pin diam, in.	Straightness tolerance, in.
1.000	0.002
0.999	0.003
0.998	0.004

If Fig. 5-1A had a specified 0.000-in. straightness tolerance at MMC, perfect straightness would be required at MMC (1.000-in. diam) and a 1.000-in.-diam gage G would be required. The 1.000-in. snap Go gage would not be required since it would then be incorporated in the straightness gage element G.

Figure 5-2A shows a 0.001-in. perpendicularity tolerance at MMC specified for a dowel pin. This tolerance must be tighter than the positional tolerance specified for the location of a pattern of such pins. If a perpendicularity tolerance is not specified, the positional tolerance includes perpendicularity. The design, in this case, requires a more restrictive perpendicularity than would be specified by the positional tolerance. The functional gage (Fig. 5-2B) for this part is a ring gage with G diam equal to 1.000 in.; the gage datum element surface must firmly contact the part datum feature A, and this must be determined by suitable mechanical means. This perpendicularity check must be performed separately on each

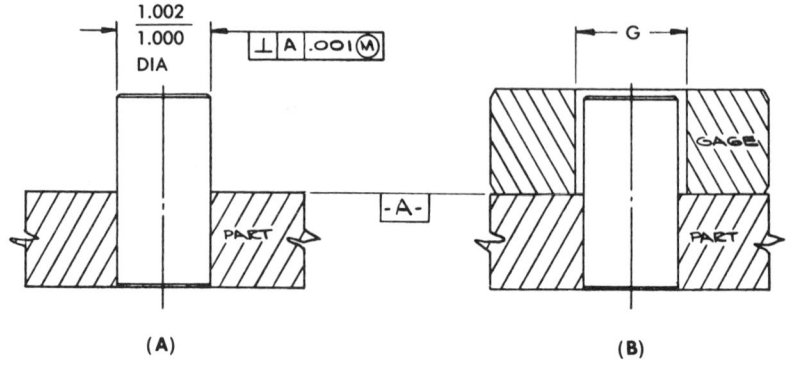

Fig. 5-2. Part and gage for perpendicularity tolerance of a dowel.

pin if several are shown in a pattern (Table 5-2). Gage diam G is determined as follows:

$$\begin{array}{r} 1.002 \text{ in., MMC size of dowel pin} \\ \text{plus } \underline{0.001} \text{ in., perpendicularity at MMC} \\ \overline{1.003} \text{ in., gage element size } G \end{array}$$

Table 5-2. Relation Between Pin Diam and Perpendicularity Tolerance

Finished pin diam, in.	Perpendicularity tolerance, in.
1.002	0.001
1.001	0.002
1.000	0.003

This part requires two ring gages: a 1.002-in. Go gage, and a 1.000-in. Not-Go gage.

If Fig. 5-2A specified perpendicularity at datum feature A of 0.000 in. at MMC, the pin would have to be absolutely perpendicular to part datum surface A at MMC and the G diam of the gage shown in Fig. 5-2B would be 1.002 inch. If the above 0.000 MMC callout were used, the 1.002-in. Go ring gage would not be required.

Figure 5-3A shows 0.005-in. perpendicularity at MMC specified for an internal feature (the 1.000–1.005-in.-diam hole) from primary part datum feature A. Design intent requires that the 1.000–1.005-in.-diam hole must be perpendicular within 0.005 in. at MMC to part datum feature A, but this tolerance may increase to 0.010 in. when the hole is 1.005-in.-diam (Table 5-3). Figure 5-3B shows the required functional gage and part. The part must bottom in this gage, and a suitable mechanical verification is required. Gage element G is determined as follows:

Chap. 5 Gaging Form Tolerances

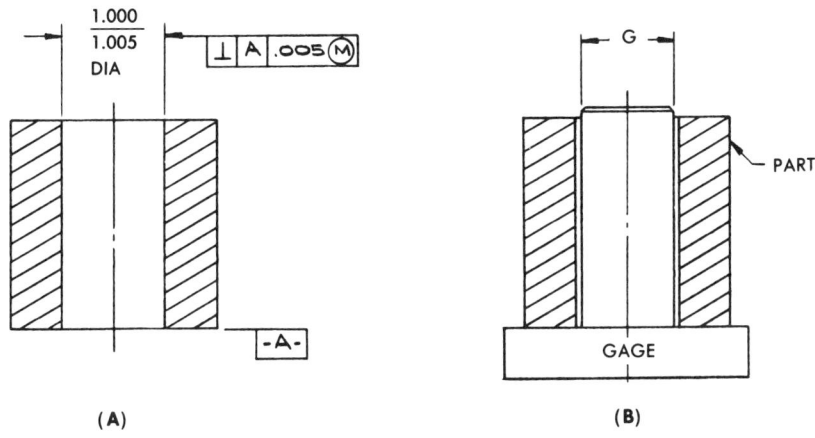

Fig. 5-3. Part and gage for perpendicularity tolerance of an internal feature.

$$\frac{\begin{array}{l} 1.000 \text{ in., MMC size of hole} \\ \text{minus } 0.005 \text{ in., perpendicularity tolerance specified at MMC} \end{array}}{0.995 \text{ in., diam } G \text{ gage element}}$$

Table 5-3. Relation Between Hole Diam and Perpendicularity Tolerance

Finished hole diam, in.	Perpendicularity tolerance, in.
1.000	0.005
1.001	0.006
1.002	0.007
1.003	0.008
1.004	0.009
1.005	0.010

Two plug gages are required: a 1.000-in. Go gage, and a 1.005-in. Not-Go gage. Had the drawing callout in Fig. 5-3A required perpendicularity of 0.000 in. to datum feature A and at MMC, the hole would have to be exactly perpendicular at MMC and the G gage element would be 1.000 inch. Since the G gage element would be a Go gage, the only plug gage required would be the 1.005-in. Not-Go gage.

Figure 5-4A specifies the perpendicularity of one hole to a pair of "in-line" part datum features. Design intent requires (1) that the 0.500–0.503-in. holes must be *in line* at MMC (when they are 0.500-in.-diam), and (2) that the 0.600–0.604-in.-diam hole must be perpendicular within 0.005 in. at MMC (when it is 0.600 in.) to the part datum features when they are 0.500 in., or at MMC. The gage datum element G (Fig. 5-4B) will be 0.500-in.-diam, the MMC size of the part datum features. Gage element G' will be 0.595-in.-diam, determined as follows:

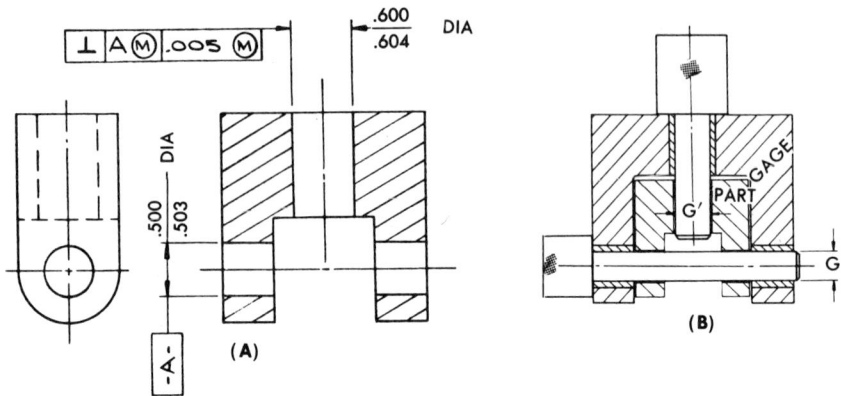

Fig. 5-4. Perpendicularity from a twin-element "datum" and related functional gage.

        0.600 in., MMC size of the perpendicular hole
minus 0.005 in., perpendicularity tolerance allowed at MMC
        $\overline{0.595}$ in. gage element G'

Three plug gages are required, as follows: (1) a 0.503-in. Not-Go gage (the 0.500-in. Go is incorporated in G gage element); (2) a 0.600-in. Go gage; and (3) a 0.604-in. Not-Go gage.

CHAPTER 6

SELECTED PORTIONS OF ANS Y14.5, DIMENSIONING AND TOLERANCING FOR ENGINEERING DRAWINGS

Extracted from ANSI Standard Drafting Practices, Dimensioning and Tolerancing for Engineering Drawings (ANS Y14.5), with the permission of the publisher, The American Society of Mechanical Engineers, United Engineering Center, 345 East 47th Street, New York, N.Y. 10017.

1. GENERAL DIMENSIONING

1.1 General. This standard establishes the rules, principles, and methods of dimensioning and tolerancing used to specify design requirements on engineering drawings. It also establishes uniform practices for stating and interpreting these requirements.

1.2 Figures. The figures given in this standard are to illustrate meanings only. In some instances figures have been over-detailed for emphasis, and in other instances figures are by intent not complete. The numerical values of dimensions and tolerances are illustrative only. Titles are included only where they add to the clarity or the intent of the figure.

1.3 Notes. Notes in capital letters are intended to appear on finished drawings. Notes in lower case letters are explanatory only and are not intended to appear on drawings.

1.4 Reference to this Standard. When drawings are based on this standard, this fact shall be noted on the drawings or in a document referenced on the drawings.

1.5 Definitions

1.5.1 Dimension. A dimension is a numerical value expressed in appropriate units of measure and indicated on a drawing along with lines, symbols, and notes, to define a geometrical characteristic of an object.

1.5.2 Basic Dimension. A dimension specified on a drawing as BASIC is a theoretical value used to describe the exact size, shape, or location of a feature. It is used as a basis from which permissible variations are established by tolerances on other dimensions or in notes.

1.5.3 Reference Dimension. A reference dimension is a dimension usually without tolerance used only for information purposes and does not govern production or inspection operations.

1.5.4 Datum. Datums are points, lines, planes, cylinders, etc. assumed to be exact for purposes of computation from which the location or geometric relationship (form) of features of a part may be established.

1.5.5 Feature. Features are specific characteristics or component portions of a part and may include one or more surfaces such as holes, screw threads, profiles, faces, or rabbets.

1.5.6 Nominal Size. The nominal size is the designation which is used for the purpose of general identification, e.g. 1.50 IPS, 0.062 stock size, etc.

1.5.7 Basic Size. The basic size is that size from which the limits of size are derived by the application of allowances and tolerances.

1.5.8 Actual Size. An actual size is a measured size.

1.5.9 Limits of Size. The limits of size are the applicable maximum and minimum sizes.

1.5.10 Maximum Material Condition (MMC). The condition where the feature contains the maximum amount of material, e.g., minimum hole diameter and maximum shaft diameter.

1.5.11 Regardless of Feature Size (RFS). The condition where tolerance of position or form must be met irrespective of where the feature lies within its size tolerance.

1.5.12 Allowance. An allowance is an intentional difference between the maximum material limits of mating parts. It is the minimum clearance (positive allowance) or maximum interference (negative allowance) between such parts.

1.5.13 Tolerance. A tolerance is the total amount by which a specific dimension may vary; thus, the tolerance is the difference between the limits.

Chap. 6 *Dimensioning and Tolerancing for Engineering Drawings*   85

1.5.14 Unilateral Tolerance. A unilateral tolerance is a tolerance in which variation is permitted only in one direction from the specified dimension.

1.5.15 Bilateral Tolerance. A bilateral tolerance is a tolerance in which variation is permitted in both directions from the specified dimension.

1.5.16 Fit. Fit is the general term used to signify the range of tightness or looseness which may result from the application of a specific combination of allowances and tolerances in the design of mating parts. Fits are of four general types: clearance, interference, transition, and line.

1.5.17 Clearance Fit. A clearance fit is one having limits of size so prescribed that a clearance always results when mating parts are assembled.

1.5.18 Interference Fit. An interference fit is one having limits of size so prescribed that an interference always results when mating parts are assembled.

1.5.19 Transition Fit. A transition fit is one having limits of size so prescribed that either a clearance or an interference may result when mating parts are assembled.

1.5.20 Line Fit. A line fit is one having limits of size so prescribed that surface contact or clearance may result when mating parts are assembled.

1.6 Fundamental Rules. Dimensioning shall conform to the following rules:

(a) Dimensioning of the geometric features of parts must convey information to define clearly the engineering intent.

(b) Each dimension must have a tolerance, either applied directly, indicated by a general note, or included in a referenced document, except those dimensions specifically labeled as REF, BSC, MAX, OR MIN.

(c) Dimensions for size, form, and location of features must be complete so that no scaling of drawings is required, and so that the intended sizes and shapes can be determined without assuming any distances.

NOTE: Undimensioned drawings (loft, printed wiring, templates, master layouts, tooling layout, etc.) prepared on stable material are excluded provided the necessary control dimensions are specified.

(d) Each dimension must be expressed clearly so that it will be interpreted in only one way.

(e) No surface, line, or point, may be located by more than one toleranced dimension in any one direction. Dimensions and related data must be given only once.

(f) Dimensions must be shown between points, lines, or surfaces which have a necessary and specific relation to each other or which control the location of other components or mating parts.

(g) Dimensions shall be selected and arranged to avoid accumulation of tolerances that may permit more than one interpretation, cause unsatisfactory mating of parts, or failure in use.

(h) Where practicable, the finished part should be defined without specifying manufacturing methods. Thus, only the diameter of a hole is given, without indication as to whether it may be drilled, reamed, punched, or made by any other operation. However, where manufacturing, processing, quality assurance, or environmental information is essential to the definition of engineering requirements, it shall be specified on the drawing, or in a document referenced on the drawing.

(i) It is permissible to identify as nonmandatory certain processing dimensions that provide for finish allowance, shrink allowance, etc., provided that the final end-product dimensions are also given on the drawing or on a higher assembly. However, where dimensions that are identified as nonmandatory occur on a drawing that represents an item of supply, the corresponding end-product dimensions shall be given on the same drawing. Nonmandatory processing dimensions shall be identified by an appropriate note, such as NONMANDATORY (MFG DATA).

(j) Dimensions must be selected to give required information directly. Dimensions should preferably be shown in true profile views and refer to visible outlines rather than to hidden lines.

. . .

2.10.3 Applicability of MMC or RFS. The need to consider whether MMC or RFS applies is limited to a feature subject to variation in size (i.e., a diameter or width), where its positional or form tolerance and datum reference applies to an axis or center plane. Only these cases, where MMC as well as RFS could be specified, are considered applicable. See Fig. 92.

2.10.4 Specifying Where MMC or RFS Applies. The drawing or a document referenced on the drawing shall state where MMC or RFS applies.

TYPE OF TOLERANCE	APPLICABILITY OF MMC AND RFS	
	FOR THE FEATURE	FOR THE DATUM REFERENCE(S)
Flatness Straightness Roundness Cylindricity	Not applicable	No datum reference
Profile of any line Profile of any surface	Not applicable	If a datum reference is necessary, RFS applies.
Perpendicularity Parallelism Angularity True Position	MMC or RFS applicable if tolerance applies to axis or center plane of considered feature; not applicable if considered feature is one plane surface.	MMC or RFS applicable if datum feature has an axis or center plane; not applicable if datum feature is one plane surface.
Runout	Not applicable	Not applicable
Concentricity Symmetry	Only RFS applicable	Only RFS applicable

FIGURE 92

Chap. 6 Dimensioning and Tolerancing for Engineering Drawings

2.11 General Rules. The general rules covered by the following paragraphs have been applied to the illustrations used in this standard.

2.11.1 General Rule Applicable to Limits of Size

Rule 1. The toleranced dimensions for the size of a feature control the form as well as the size. The basic interpretation of this implied control of form is as follows:

a) No element of the actual feature shall extend beyond the envelope of perfect form at MMC. This envelope is the true form implied by the drawing.

b) The measured dimensions of the feature at any cross-section shall not be less than the minimum limit of size of an external feature nor greater than the maximum limit of size of an internal feature.

Figure 93 illustrates the extreme variations of form that are permitted by this interpretation. The MMC is indicated by phantom lines.

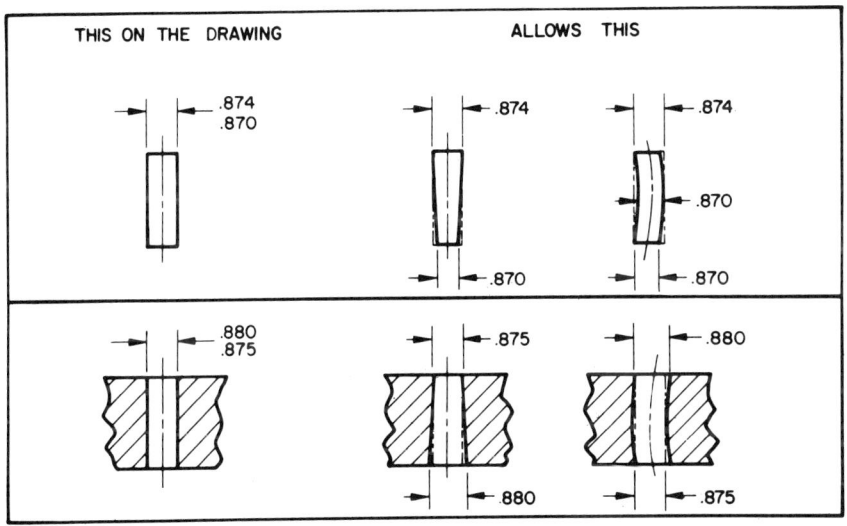

FIGURE 93 – EXTREME VARIATIONS OF FORM ALLOWED BY SIZE TOLERANCES

Note that the above stated interpretation prescribing an envelope of perfect form at MMC applies only to individual features and not to the interrelationship of features. Such interrelationship should be controlled by form or positional tolerances specified on the drawing or in a referenced document to the extent dictated by the design requirements. This control may take the form of a requirement that the above interpretation for an individual feature applies also to the interrelationship of features (i.e., requiring perfect form for the complete part at MMC).

2.11.1.1 Exceptions.

(a) INTERPRETATION FOR COMMERCIAL STOCK. This interpretation does not apply to commercial stock such as bars, sheets and tubing where established industry standards prescribe straightness, flatness, and other conditions. Where variations of form are not given on drawings of parts made from these materials, industry standards for the materials govern the surfaces that remain in the "as furnished" commercial condition for the finished part.

(b) OTHER FEATURES: Where it is desired to permit a tolerance of form to exceed the MMC-perfect-form envelope, this may be done by adding to the drawing the suitable form tolerance and a note specifically exempting the pertinent size dimension(s) from the perfect form requirement. A suitable note might read: PERFECT FORM NOT REQD AT MMC.

2.11.2 General Rules Applicable to MMC and RFS. Where reference is made to this standard and the MMC or RFS modifiers are not specified on a drawing with respect to an individual tolerance, datum reference, or both, the following rules apply:

Rule 2. True position tolerances and related datum references apply at MMC.

Rule 3. Angularity, parallelism, perpendicularity, concentricity, and symmetry tolerances, including related datum references, apply RFS. No element of the actual feature shall extend beyond the envelope of perfect form at MMC.

Note: The above rules are not applicable where a feature or datum is one plane surface. See Paragraph 2.10.3.

2.11.3 Optional Method in Lieu of Rules 2 and 3. The indication of whether MMC or RFS applies may be expressed in connection with each true position or form tolerance and each datum reference by always specifying the appropriate modifier.

2.11.4 General Rule Applicable to Screw Threads. The following rule is applicable to all symbols and notes specifying tolerances of form or position involving screw threads as toleranced features or datums:

Rule 4. Where tolerances of form or position are expressed by symbols and notes, each such tolerance applicable to a screw thread and each datum reference to a screw thread shall be understood to apply to the pitch diameter. If design requirements necessitate an exception to this general rule, a qualifying notation shall supplement the symbol or note, e.g., MAJOR DIA. In the case of symbol application, the qualifying notation shall be shown beneath the feature control symbol where applicable to the feature, and beneath the datum identifying symbol where applicable to the datum.

Chap. 6 *Dimensioning and Tolerancing for Engineering Drawings* 89

2.11.4.1 Gears and Splines. For gears and splines, a qualifying notation must be added to the symbol or note, (e.g., MAJOR DIA, MINOR DIA, PITCH DIA or PD).

3. SYMBOLS FOR TOLERANCES OF POSITION AND FORM

3.1 General. When symbols are used in lieu of notes in specifying positional and form tolerances, the symbols shall be those shown herein.

3.2 Use of Notes to Supplement Symbols. Situations may arise in which the precise geometric requirement desired is not conveyed by any symbol provided herein. In each such instance, a note should be used, either separately or supplementing a symbol, to describe the precise requirement. Symbols should be used elsewhere on the same drawing to the extent they are applicable.

3.3 Individual Symbols

3.3.1 Geometric Characteristic Symbols. The symbols denoting geometric characteristics are shown in Fig. 94.

GEOMETRIC CHARACTERISTIC SYMBOLS			
	Characteristic		Symbol
Form Tolerances	For Single Feature	FLATNESS	▱
		STRAIGHTNESS	—
		ROUNDNESS (CIRCULARITY)	○
		CYLINDRICITY	⌀
		PROFILE OF ANY LINE [1]	⌒
		PROFILE OF ANY SURFACE [1]	⌒
	For Related Features	PARALLELISM [2]	∥
		PERPENDICULARITY (SQUARENESS)	⊥
		ANGULARITY	∠
		RUNOUT [3]	↗
Positional Tolerances		TRUE POSITION	⊕
		CONCENTRICITY [4]	⊚
		SYMMETRY [5]	≡

[1] Although included under "Form Tolerances," profile tolerances control size as well as form.
[2] Parallel lines may be shown oblique.
[3] Although included under "Form Tolerances," a runout tolerance controls position as well as form.
[4] Where concentricity RFS applies, it is preferred that the runout symbol be used. Refer to Paragraph 4.12. Where concentricity at MMC applies, it is preferred that the true position symbol be used. Refer to Paragraph 4.12. Optionally, the inner circle of the symbol may be filled solid.
[5] Where symmetry applies, it is preferred that the true position symbol be used. Refer to Paragraph 4.13.

FIGURE 94

3.3.2 Datum Reference Letters. Any letter of the alphabet except "I," "O" and "Q" may be used as a datum reference letter. Each datum feature requiring identification should be assigned a different reference letter. When datum features requiring identification on a drawing are so numerous as to exhaust the single letter series, the double letter series, i.e., AA through AZ may be used.

3.3.3 Datum Identifying Symbol. The datum identifying symbol shall consist of a frame containing the datum reference letter, the latter preceded and followed by a dash. See Fig. 95. The symbol shall be associated with the datum feature by one of the methods prescribed in Paragraph 3.5 for feature control symbols. Each datum requiring identification on a drawing shall be assigned a different reference letter(s).

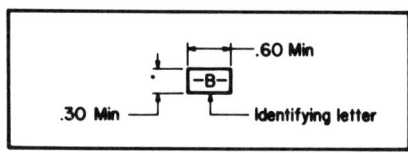

FIGURE 95 – DATUM IDENTIFYING SYMBOL

3.3.4 Symbol for Basic or True Position Dimension. The symbolic means of labeling a basic or true position dimension is by enclosing each such dimension in a frame, as shown by Fig. 96.

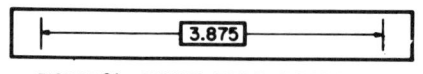

FIGURE 96 – SYMBOL FOR BASIC OR TRUE POSITION DIMENSION

3.3.5 Symbols for MMC and RFS. The symbols Ⓜ and Ⓢ are used to designate "Maximum Material Condition" and "Regardless of Feature Size," respectively. See Fig. 97. These symbols are restricted to use as modifiers in feature control symbols. The abbreviations MMC and RFS, or the spelled-out terms, are used as applicable, in notes.

SYMBOLS FOR MMC AND RFS	
Modifier	Symbol
(MMC) MAXIMUM MATERIAL CONDITION	Ⓜ
(RFS) REGARDLESS OF FEATURE SIZE	Ⓢ

FIGURE 97

3.4 Combined Symbols

3.4.1 Feature Control Symbol. A positional or form tolerance shall be stated by means of the feature control symbol which shall consist of a frame containing the geometric characteristic symbol followed by the permissible tolerance, and in some cases by the modifier Ⓜ or Ⓢ. See Paragraphs 2.11.2 and 3.3.5. A vertical line shall separate the symbol from the tolerance. See Fig. 98.

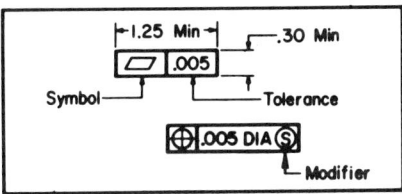

FIGURE 98 – FEATURE CONTROL SYMBOLS

3.4.1.1 Feature Control Symbol Incorporating Datum References. Where a positional or form tolerance must be related to a datum(s), this relationship shall be stated in the feature control symbol by placing the datum reference letter(s) between the geometric characteristic symbol and the tolerance. Vertical lines shall separate these entries. The length of the frame shall be increased as necessary to avoid crowding. See Figs. 99, 100, and 101.

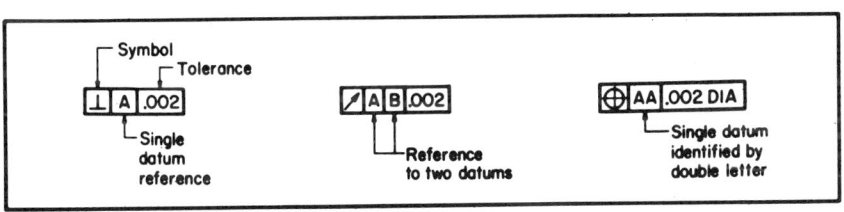

FIGURE 99 – FEATURE CONTROL SYMBOLS INCORPORATING DATUM REFERENCES

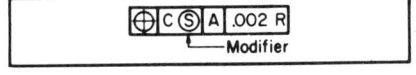

FIGURE 100 – DATUM REFERENCES SHOWING ORDER OF PRECEDENCE

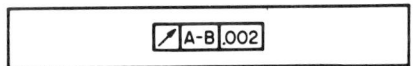

FIGURE 101 – MULTIPLE DATUM FEATURES ESTABLISHING SINGLE REFERENCE

3.4.2 Datum References Showing Order of Precedence. Each datum reference letter (with Ⓜ or Ⓢ modifier, if required) shall be entered in the desired order of precedence, from left to right, in the feature control symbol. Thus datum reference letter entries may not be in alphabetical order. See Fig. 100. For further detail information on datums see Appendix A.

3.4.2.1 Multiple Datum Features Establishing a Single Datum Reference. Where a single datum reference is established by multiple datum features, the datum reference letters shall be entered in a single compartment, a dash being placed between the letters. See Fig. 101.

3.4.3 Combined Feature Control and Datum Identifying Symbol. Where a feature serves as a datum and also is controlled by a positional or form tolerance, the feature control symbol and the datum identifying symbol shall be combined, as shown in Fig. 102. In such cases, the length of the frame for the datum identifying symbol may be either the same as that of the feature control symbol or .60 inch minimum.

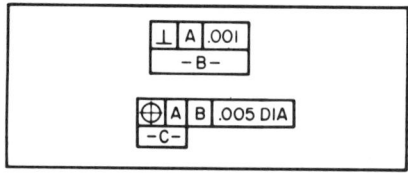

FIGURE 102 – COMBINED FEATURE CONTROL AND DATUM IDENTIFYING SYMBOLS

3.5 Symbol Placement. The feature control symbol shall be associated with the feature(s) being toleranced by one of the following methods, as depicted in Fig. 103:

a) Adding the symbol to a note or dimension pertaining to the feature(s).

b) Running a leaderline from the feature(s) to the symbol.

c) Attaching a side, end, or corner of the control symbol frame to an extension line from the feature.

d) Attaching a side or end of the frame to the dimension line pertaining to the feature.

3.6 Identification of Tolerance Zone. In the symbolic expression terms such as DIA, TOTAL, WIDE ZONE, FIR, or ON RAD. need not be included in the feature control symbol except in the case of true position, concentricity and symmetry tolerances where DIA or TOTAL shall be specified and shall represent the diameter of a cylindrical zone or the distance between two parallel zone planes as applicable. Where it is desired to identify a true position tolerance as a radius the modifier "R" shall be specified and shall represent the radius of a cylindrical zone or one half the distance between two parallel zone planes as applicable. For round-

Chap. 6 Dimensioning and Tolerancing for Engineering Drawings

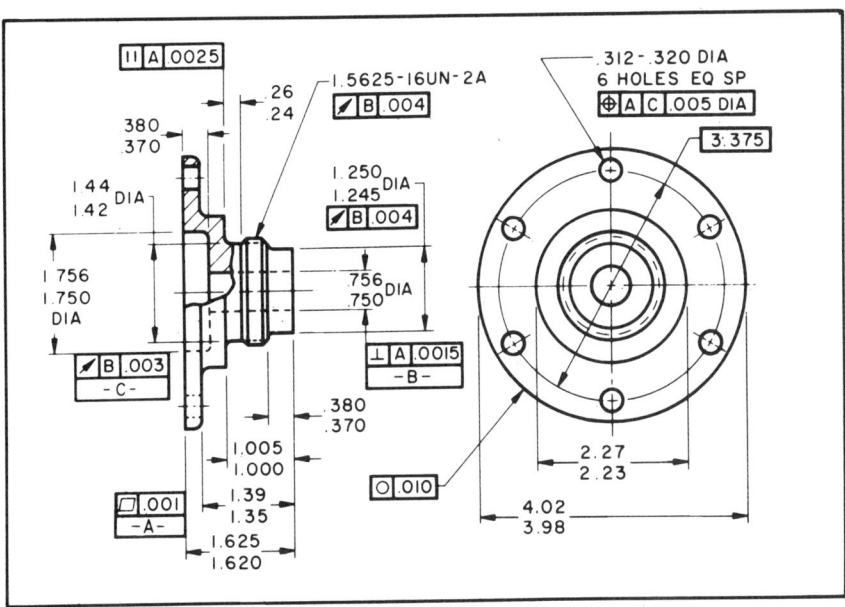

FIGURE 103 – APPLICATION OF GEOMETRIC CHARACTERISTIC SYMBOLS

ness, cylindricity, or runout, the tolerance value shown shall represent the full indicator reading obtained by rotating the part on its axis.

3.7 Former Symbols. The symbol ∼ formerly was prescribed for use in expressing a tolerance on flatness. The symbols ⌒ and — formerly were prescribed for use in expressing tolerances on straightness and flatness. Whenever these symbols appear on existent drawings, they shall be interpreted as governing that geometric characteristic which applies to the specific case shown.

4. TOLERANCES OF POSITION

4.1 General. A positional tolerance is the tolerance assigned to a dimension that locates one or more features in relation to another feature. Typical feature relationships where positional tolerances apply include the following:

a) Center distance between features such as holes, slots, or bosses.

b) Location of features such as holes, slots, bosses or fixed dowels from datum features such as faces or rabbets.

c) Concentricity between the diameters of a multidiameter part (where concentricity is the desired condition and the tolerance gives the permissible eccentricity).

4.2 Tolerance Zones. A true position tolerance and a concentricity tolerance both specify tolerance zones within which the considered feature, its axis, or center plane must be contained. For a description of these tolerance zones, see Fig. 104.

GEOMETRIC DESCRIPTION OF TOLERANCE ZONES

TOLERANCE			INTERPRETATION			
Condition (Type of tolerance)	Toleranced feature	Datum feature (When applicable)	Specified tolerance value represents	Description of tolerance zone	Datum relationship (When applicable)	Example
True position	Cylindrical surface	Plane or cylindrical surfaces	Diameter of a cylindrical zone. (For the diameter method) Radius of a cylindrical zone. (For the radius method)	Space within a cylindrical zone	Zone axis is coincident with true position and oriented in relation to the specified or implied datums	Figure 122
	Median plane of non-cylindrical features	Plane or cylindrical surfaces	Distance between two parallel planes. (For the diameter method) Distance either side of the center plane. (For the radius method)	Space between two parallel planes	Zone center plane is coincident with true position and oriented in relation to the specified or implied datums	Figure 151
Concentricity	Surface of revolution	Cylindrical, conical, or plane surfaces used to establish a datum axis	Diameter of a cylindrical zone	Space within a cylindrical zone	Zone axis is concentric with the datum axis	Fig. 153

FIGURE 104

4.12 Coaxiality Controls. Where two or more surfaces of revolution such as cylinders, spheres, cones, etc. are generated about a common axis, the amount of permissible deviation from such coaxiality may be expressed directly (by a concentricity tolerance), or indirectly (either by a runout tolerance or by a true position tolerance). Selection of the proper control depends on the nature of the functional requirement as explained in the following recommendations:

a) Concentricity Tolerance. Concentricity is the condition of surfaces of revolution wherein they have a common axis. Concentricity tolerances are expressed and interpreted as shown in Fig. 153.

Note. Irregularities in the form of the feature to be inspected may make it difficult to actually establish the axis of the feature. For instance, a

Chap. 6 Dimensioning and Tolerancing for Engineering Drawings 95

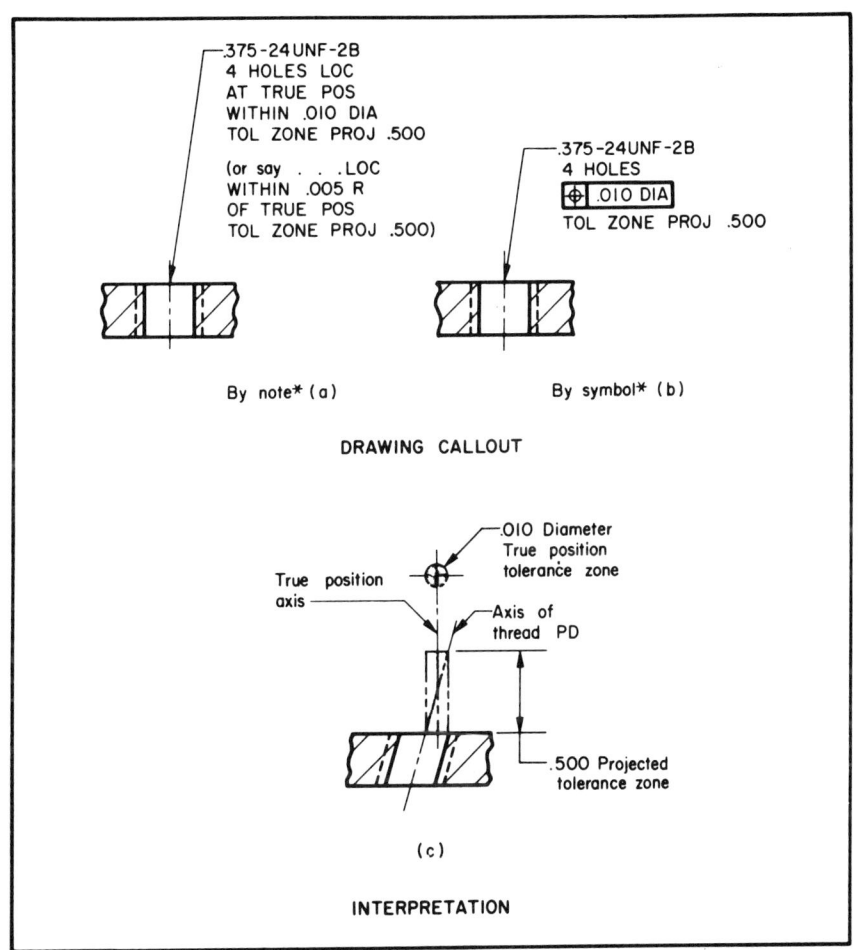

FIGURE 139 - PROJECTED TOLERANCE ZONE CALLOUT AND INTERPRETATION

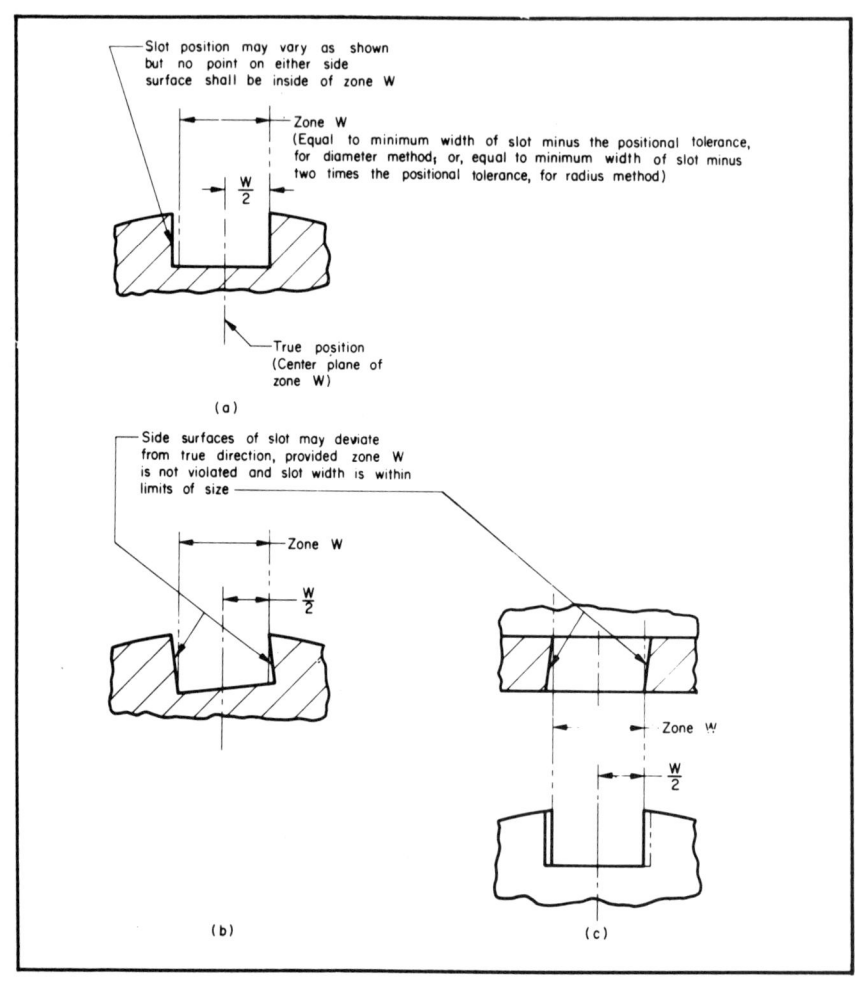

FIGURE 150 – TOLERANCE ZONE FOR SURFACES OF SLOT AT MMC

Chap. 6 *Dimensioning and Tolerancing for Engineering Drawings*   97

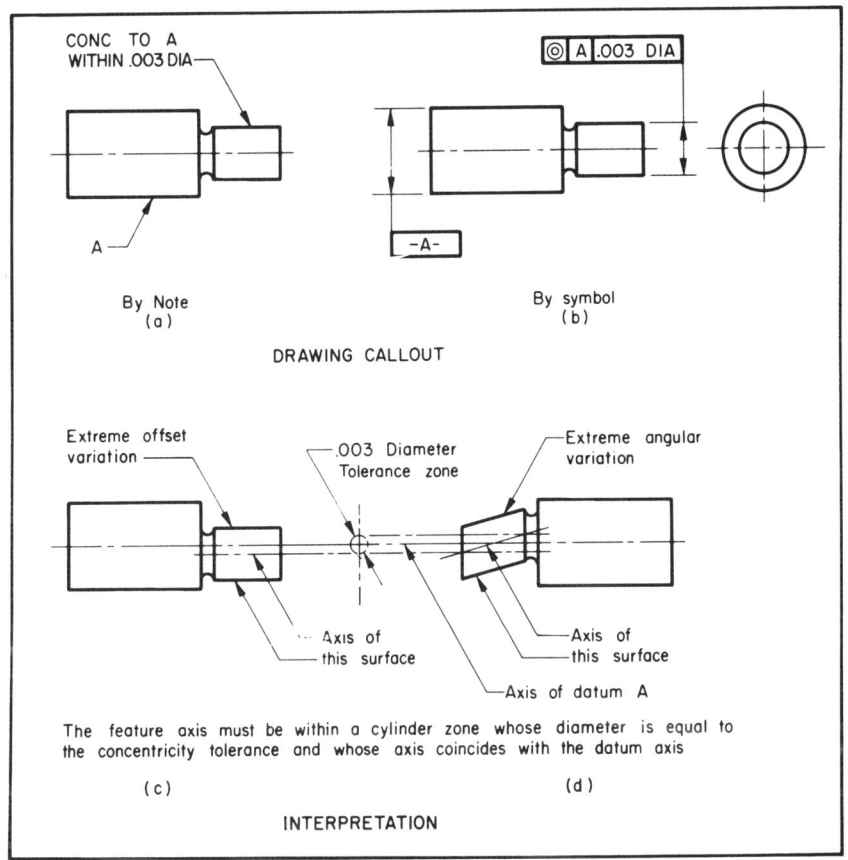

FIGURE 153 - CONCENTRICITY CALLOUT AND INTERPRETATION

nominally cylindrical surface may be bowed or out-of-round in addition to being offset from its datum feature; in such cases, finding the axis of the feature may entail a time-consuming analysis of the surface. Therefore, unless there is a definite need for the control of axes (as in the case shown in Fig. 154), it is recommended that the control be specified in terms of runout tolerance (see b) or true position tolerance (see c), below.

b) *Runout Tolerance*. Where a part has nominally coaxial features and the permissible deviations (from this condition) at the surface of such features must be controlled to a given amount regardless of whether the features are at MMC, the use of a runout control is recommended; see Paragraph 5.6.5. It is further recommended that, where applicable, the runout tolerance be modified by the word "circular"; this permits a less strict control and a simpler inspection (see Paragraph 5.6.5.5).

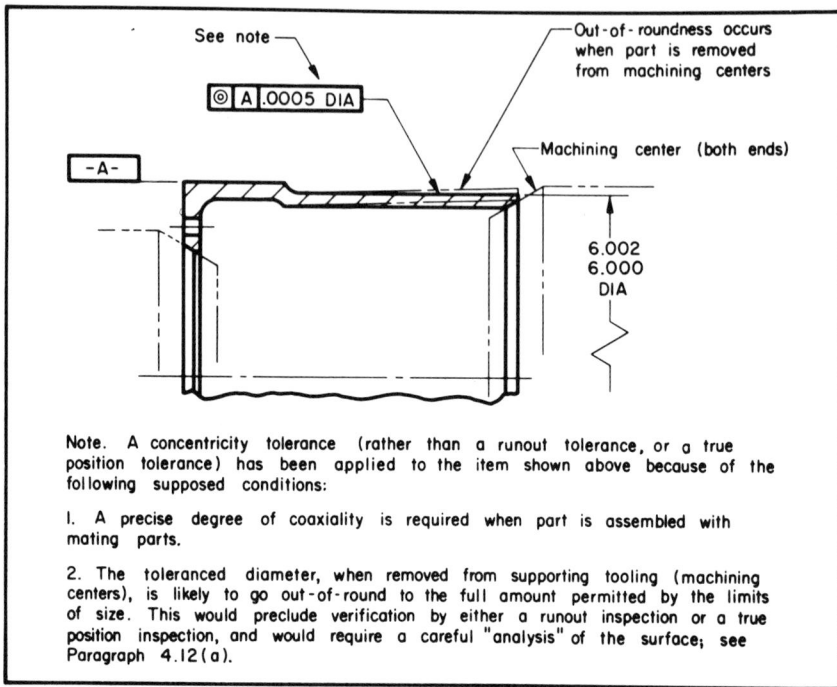

FIGURE 154 – EXAMPLE WHERE CONCENTRICITY TOLERANCE IS REQUIRED

c) True Position Tolerance. Where it is necessary to control a nominally coaxial relationship between two features, merely to assure assemblability, the use of a true position tolerance is recommended; see Figs. 155 and 157. The tolerance for the feature should be given on an MMC basis while, for the datum feature, the tolerance may be given either on an MMC or an RFS basis. Where true position tolerancing is applied for control of coaxiality, it is subject to the same interpretation principles given in Paragraph 4.5.1.2.

Note. A true position control of this kind usually permits (but does not dictate) the use of a simple receiver gage, for inspection. Fig. 156 illustrates the application of a receiver gage for the head pin described in Fig. 155. In (a) of Fig. 156, both the body and the head of the actual pin are at MMC; in (b), only the body is at MMC; in (c), neither head nor body is at MMC.

4.13 Symmetry. Symmetry is a condition wherein a part or a feature has the same contour and size on opposite sides of a central plane, or a condition in which a feature is symmetrically disposed about the central plane of a datum feature.

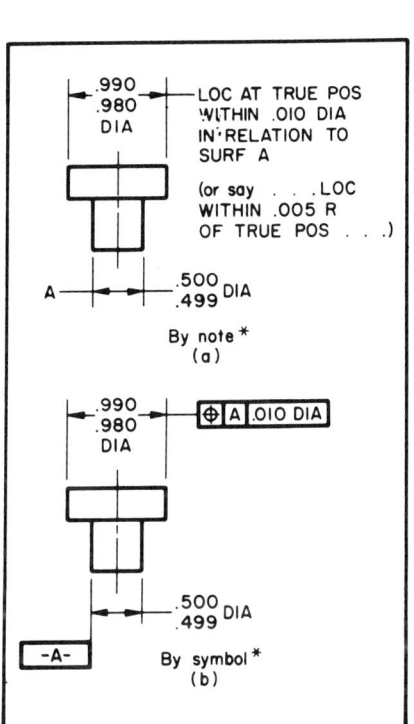

FIGURE 155 – TRUE POSITION TOLERANCING FOR COAXIALITY

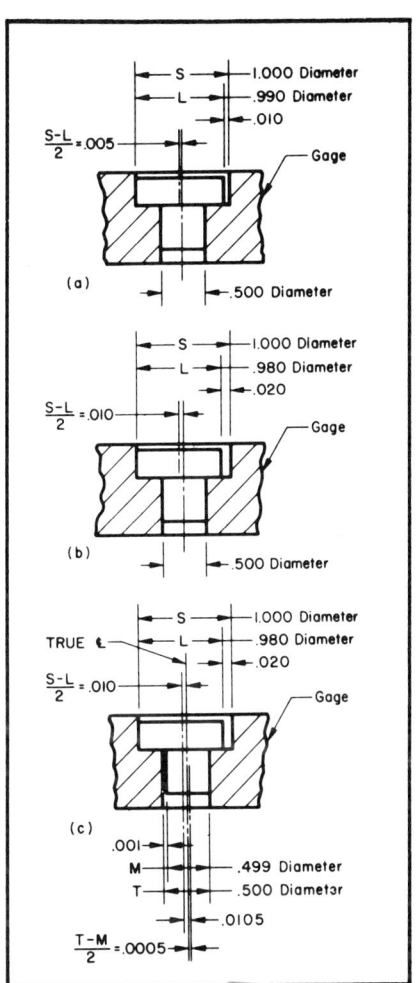

FIGURE 156 – VARIOUS CONDITIONS OF PART SHOWN IN FIGURE 155, AND MATING HOLE

4.13.1 Symmetry Tolerance. Where it is required that a feature be located symmetrically with respect to a datum feature, the use of true position tolerancing is recommended; this permits the tolerance to be expressed on an MMC basis (as shown in Fig. 158) or on an RFS basis (as shown in Fig. 159). If it is preferred to use a symmetry tolerance, the method shown in Fig. 160 may be followed. However, it will be seen from the interpretation in this figure that a symmetry tolerance, when so expressed, always applies on an RFS basis.

100   Dimensioning and Tolerancing for Engineering Drawings   Chap. 6

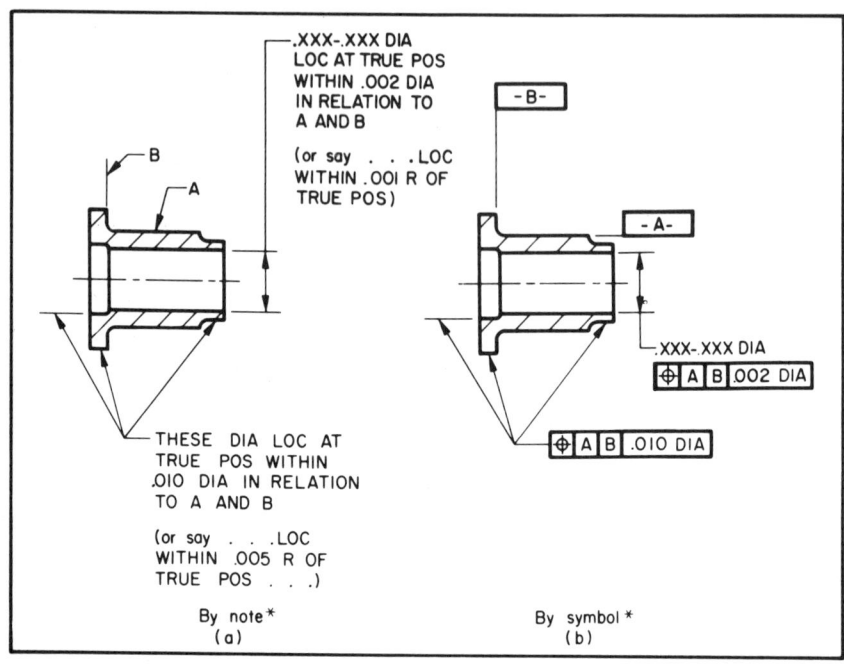

FIGURE 157 – TRUE POSITION TOLERANCING FOR COAXIALITY

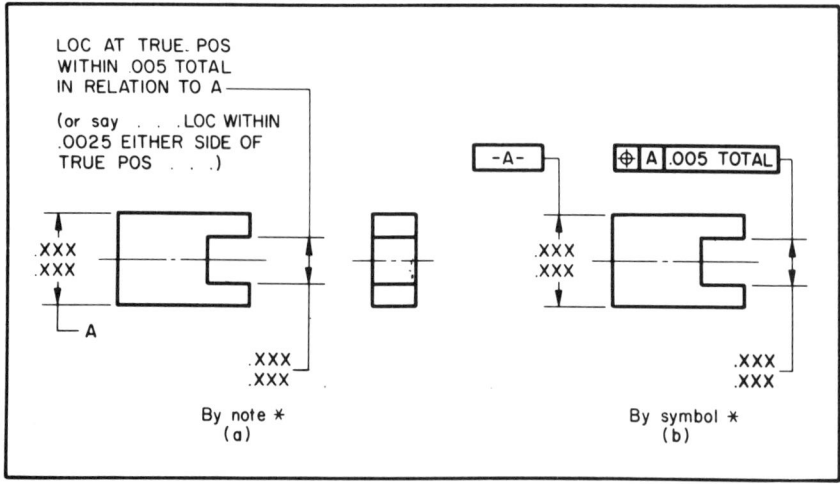

FIGURE 158 – TRUE POSTION TOLERANCING FOR SYMMETRY

Chap. 6 Dimensioning and Tolerancing for Engineering Drawings   101

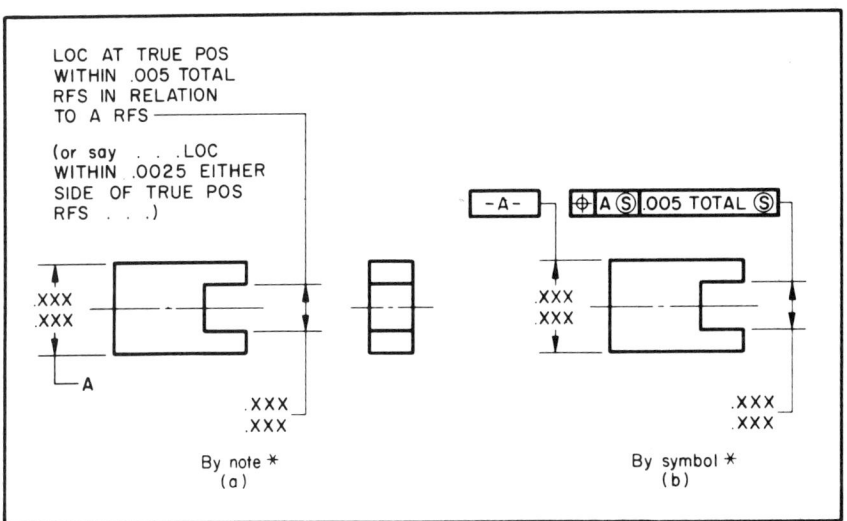

FIGURE 159 – TRUE POSITION TOLERANCING FOR SYMMETRY, RFS

5. TOLERANCES OF FORM

5.1 General. This section establishes the rules, principles and methods of dimensioning and tolerancing applicable to the control of form for the various geometrical shapes, and the control of free-state variations.

5.2 Form Control. Form tolerances control more specifically the conditions of straightness, flatness, roundness, cylindricity, profile of any surface or line, angularity, parallelism, perpendicularity, and runout. Since tolerances of size and location control form to a certain degree, the extent of this control should be clearly understood before specifying form tolerances. Furthermore, consideration should be given to established workshop practices which may be sufficiently reliable to provide the required accuracy, thereby obviating the expression of certain form tolerances on drawings. It may be desirable to control form to a practical degree by referring to a separate document which establishes limits of good workmanship.

5.3 Specifying Form Tolerance. Form tolerances are specified by drawing notes, symbols, or supplementary documents. In the following paragraphs, both the symbolic and note callouts are illustrated, along with the interpretation of the tolerance. These tolerances specify the maximum permissible variations from the desired form (within the limits of size) and apply to all points on the considered surface or line, unless otherwise specified. Where applicable, the examples illustrate the relationship of tolerances of size and position to form tolerances if the terms regardless of feature size

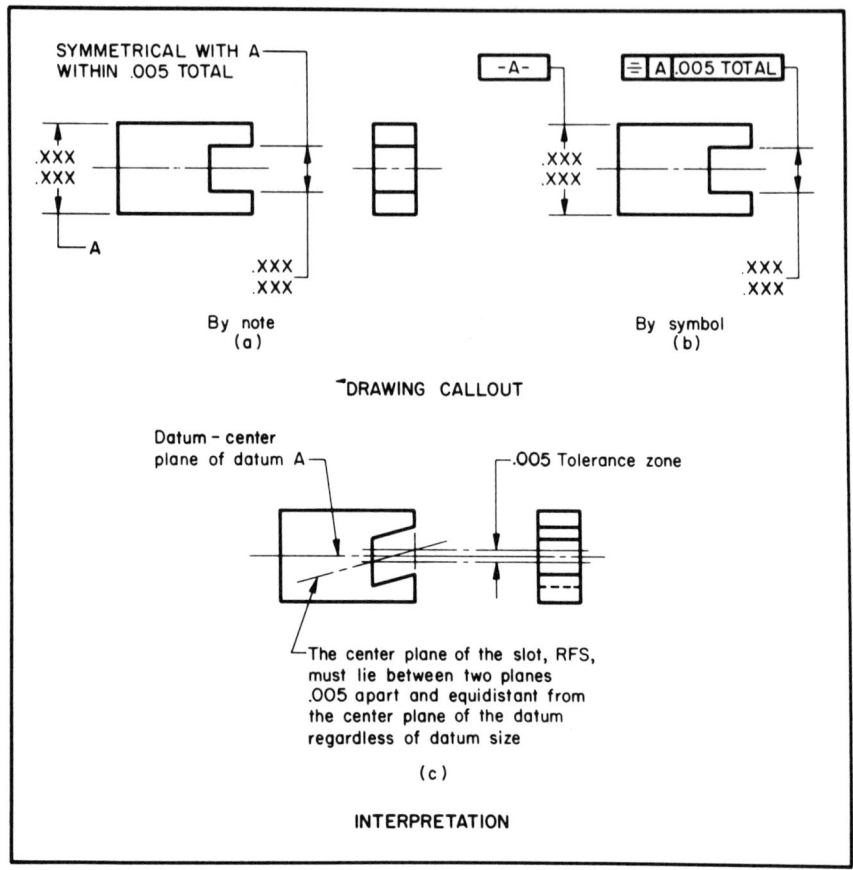

FIGURE 160 - SYMMETRY CALLOUT AND INTERPRETATION

(RFS), or maximum material condition (MMC) are applied. Since a more liberal form tolerance may be allowed where the MMC modifier is applied, careful attention must be given in applying these terms to ensure design intent is fulfilled. For the general rules applicable to the limits of size and the application of MMC and RFS to form tolerances, see Par. 2.11.

5.3.1 Form Tolerance Zones. A form tolerance specifies a tolerance zone within which the considered feature, its axis, or center plane must be contained. Depending upon the geometrical shape of the considered feature and the manner in which it is dimensioned, the tolerance zone may be one of those described in Fig. 161.

5.3.1.1 Application of the Modifier, Total. If the form tolerance value is specified in a note, it shall be followed by the modifier, TOTAL except when roundness, cylindricity, or runout is used. For these exceptions, the

Chap. 6   Dimensioning and Tolerancing for Engineering Drawings   103

TOLERANCE			INTERPRETATION			
Condition (Type of tolerance)	Toleranced feature	Datum feature (If applicable)	Specified tolerance value represents	Description of tolerance zone	Datum relationship (If applicable)	Example
Straightness	Straight line element of a surface	None	Distance between two parallel straight lines	Area between two parallel straight lines		Fig. 162
Flatness	Plane surface	None	Distance between two parallel planes	Space between two parallel planes		Fig. 163
Roundness	Circular element of a surface of revolution	None	Distance between two coplanar, concentric circles	Area between two coplanar, concentric circles		Fig. 165
Cylindricity	Cylindrical surface	None	Distance between two coaxial cylinders measured radially	Space between two coaxial cylinders		Fig. 167
Profile[1]	Any surface	None	Distance between two uniform boundaries measured along a line normal to the basic profile at all points	Space between two uniform boundaries having the same basic shape as the considered feature		Fig. 169
	Any line			Area between two uniform boundaries having the same basic shape as the considered feature		Fig. 171
Angularity	Plane surface	Plane or cylindrical surface	Distance between two parallel planes	Space between two parallel planes	Zone is at the specified angle with the datum plane or axis	Fig. 172
Parallelism	Plane surface	Plane surface	Distance between two parallel planes	Space between two parallel planes	Zone is parallel to the datum plane	Fig. 173
	Cylindrical surface	Plane surface				Fig. 174
	Cylindrical surface	Cylindrical surface	Diameter of a cylindrical zone	Space within a cylindrical zone	Zone axis is parallel to the datum axis	Fig. 176

[1] Profile tolerance is included in the general category of form tolerances. However, a profile tolerance may control size as well as form.

FIGURE 161 (SHEET 1) – GEOMETRIC DESCRIPTION OF TOLERANCE ZONES

TOLERANCE				INTERPRETATION			
Condition (Type of tolerance)	Toleranced feature	Datum feature (If applicable)	Specified tolerance value represents	Description of tolerance zone	Datum relationship (If applicable)	Example	
Perpendicularity	Plane surface	Plane surface	Distance between two parallel planes	Space between two parallel planes	Zone is perpendicular to the datum plane	Fig. 178	
	Median plane of a noncylindrical feature	Plane surface				Fig. 179	
	Cylindrical surface	Cylindrical surface	Diameter of a cylindrical zone	Space within a cylindrical zone	Zone axis is perpendicular to the datum plane	Fig. 181	
	Cylindrical surface	Plane surface				Fig. 182	
	Radial element of a surface	Cylindrical surface	Distance between two parallel, straight lines	Area between two parallel, straight lines	Zone is perpendicular to and coplanar with the datum axis	Fig. 189	
Runout	Surface of revolution having longitudinal elements which are straight or circular	Cylindrical, conical, or plane surfaces used to establish a datum axis	Full indicator reading as the surface is traversed, or the distance between two coaxial shapes measured along a line normal to the considered surface and passing through the datum axis	Space between two coaxial shapes having the same geometrical shape as the considered feature	Zone is concentric with the datum axis	Fig. 193	
	Plane surface of revolution	Cylindrical, conical, or plane surfaces used to establish a datum axis	Full indicator reading as the surface is traversed, or the distance between two parallel planes	Space between two parallel planes	Zone is perpendicular to the datum axis	Fig. 193	
	Circular element of a surface of revolution (includes an end face of a surface of revolution)		Maximum full indicator reading in one revolution	Varies	Zone is concentric with the datum axis	Fig. 199 & 200	

FIGURE 161 (SHEET 2) – GEOMETRIC DESCRIPTION OF TOLERANCE ZONES (CONT'D)

modifier, ON R, shall be used for roundness and cylindricity, and the modifier, FIR, shall be used for runout. If the form tolerance value is specified in a feature control symbol, no modifier is required.

5.3.1.2 Form Tolerance Zones For Limited Area or Length. Some designs only require the control of form over a limited area or length of the surface, rather than the customary control over the total surface. In these instances, the specific area, or length, and its location are specified with the appropriate dimensions.

5.3.1.3 Form Tolerance Zones on a Unit Basis. It may be desirable to control the form of a surface on a unit basis in terms of inches per inch by one of the following methods:

Method 1. By specifying the unit variation value and the total variation value, the interpretation shall be as follows:

The unit variation value means that any one-inch portion of the considered surface must lie within the unit value specified, and that all portions so measured must lie within the boundaries of the total variation tolerance. Typical drawing notes are:

STR WITHIN .005
TOTAL AND WITHIN
.001/IN.

FLAT WITHIN .005
TOTAL AND WITHIN
.001/IN.

PERP TO DATUM A
WITHIN .005 TOTAL
AND WITHIN .001/IN.

Method 2. By specifying the unit variation value only, the interpretation shall be as follows:

The unit variation value establishes the maximum change per unit of length with the total variation controlled only by the number of unit lengths in the surface. Typical drawing notes are:

STR WITHIN .001/IN.

FLAT WITHIN .001/IN.

PERP TO DATUM A
WITHIN .001/IN.

5.3.2 Symbols. For the definition, description, and method of applying the feature control symbol and form tolerance symbols, see Paragraphs 3 thru 3.7.

5.3.3 Liberalized Form Tolerance by Applying MMC. The application of MMC to a form tolerance means that the tolerance applies at MMC and is increased as the feature(s) depart(s) from MMC by the amount of such departure. Therefore, if the control of form is required only at MMC, a liberalized form tolerance is permitted. The method of indicating on the drawing that the liberal interpretation is permitted is to add AT MMC to the note or symbolically in the feature control symbol.

5.3.3.1 Application of Zero Tolerance at MMC. Where variations of form (perpendicularity, etc.), pertaining to the axis or center plane of a feature, are to be contained within the maximum size limit of an external (male) feature or the minimum size limit of an internal (female) feature, the notation, PERP TO DATUM A, ZERO AT MMC should be used. Such a notation means that if the features are finished everywhere on their maximum material limits of size, they must be perfect in form (i.e., truly perpendicular) with respect to the datum. Variations of form are only permissible if the features are away from their maximum material limits of size provided that the minimum material limits of size are observed.

5.4 Application of Form Tolerances. Form tolerances should be specified for all features critical to function and interchangeability:

a) Where established workshop practices cannot be relied upon to provide the required accuracy.

b) Where documents establishing suitable standards of workmanship cannot be prescribed.

c) Where the tolerances of size and location do not provide the necessary control.

5.5 Form Tolerances for Single Features

5.5.1 Measurements From Primary Reference Standards. To ensure that design intent will be met, the interpretation of form tolerances for single features should be clear to manufacturing personnel and to inspection when verifying conformance. In verifying straightness, flatness, roundness, cylindricity, and profile, the considered feature is measured in relationship to a primary reference standard simulated by inspection and manufacturing equipment. For instance, in measuring the flatness of a surface, one boundary of the form tolerance zone is established by a primary reference standard (such as a surface plate) touching the high points of the considered feature. The other boundary for the zone is located from this primary reference standard at a distance equal to the specified flatness tolerance. In some instances, the considered feature will not bottom easily against the primary reference standard, but has a tendency to "teetertotter." For the purposes of determining conformance, the "teeter-totter" effect should be equalized to the extent that it will minimize the surface inaccuracies of the considered feature with respect to the primary reference

Chap. 6 *Dimensioning and Tolerancing for Engineering Drawings* 107

standard. In measuring cylindricity, one boundary of the form tolerance zone is established by an enveloping reference cylinder just touching the high points of the considered feature. The other boundary for the zone is concentric with and established from the reference cylinder at a distance equal to the cylindricity tolerance. In lieu of an actual external or internal reference cylinder for establishing effective size and coincident axes, the primary reference standard is simulated by inspection or manufacturing methods and equipment.

5.5.2 Straightness Tolerance. Straightness is a condition where an element of a surface is a straight line. A straightness tolerance specifies a tolerance zone of uniform width along a straight line, within which all points of the considered line must lie. In specifying a straightness tolerance, the leader shall run from the symbol or note to the surface, and applied in a view where the surface elements to be controlled are represented as a straight line. Fig. 162 shows a means of specifying the permissible variation.

5.5.3 Flatness Tolerance. Flatness is the condition of a surface having all elements in one plane. A flatness tolerance specifies a tolerance zone confined by two parallel planes within which the surface must lie. Fig. 163 shows the method of specifying the permissible variation. The expression

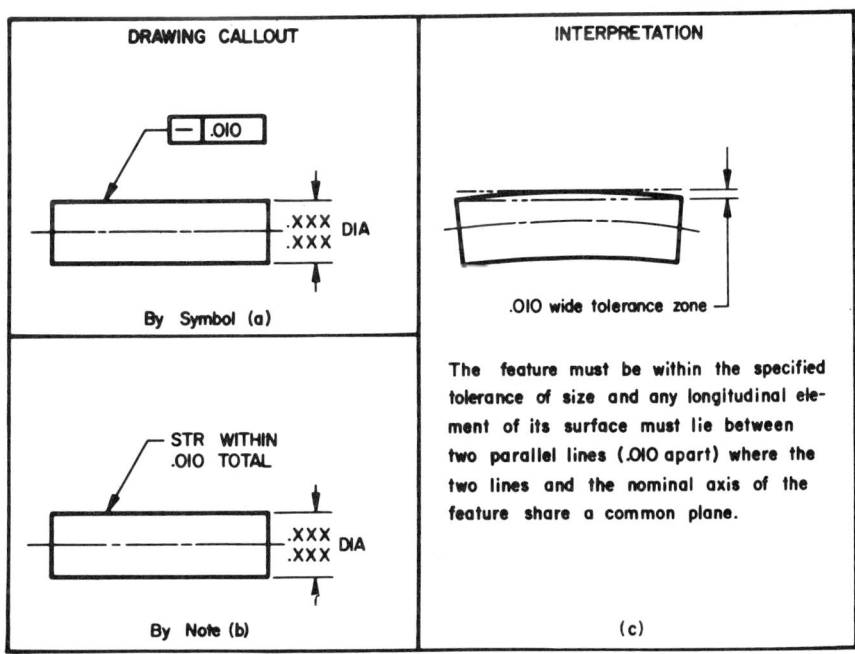

FIGURE 162 – SPECIFYING STRAIGHTNESS

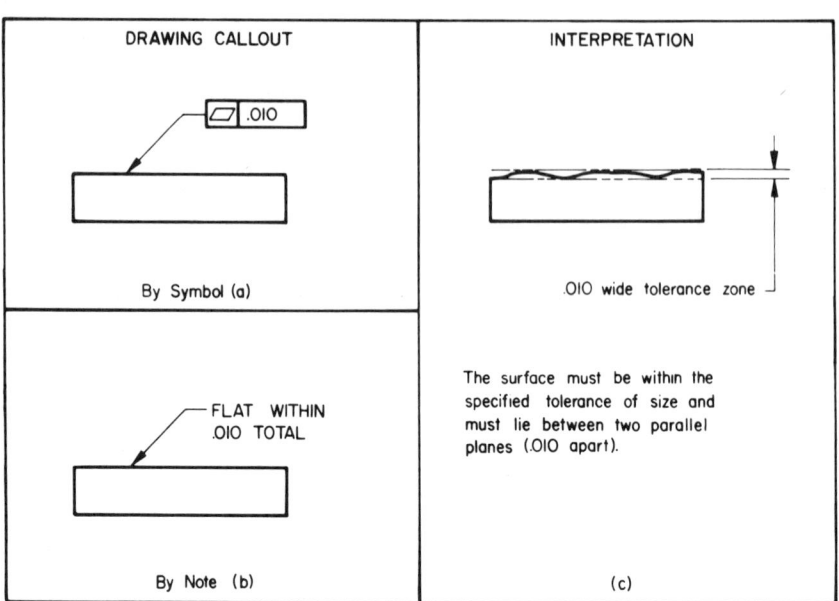

FIGURE 163 – SPECIFYING FLATNESS

MUST NOT BE CONCAVE or MUST NOT BE CONVEX may be added if necessary.

5.5.4 Roundness Tolerance. Roundness is a condition of a surface of revolution such as a cylinder, cone, or sphere, where all points of the surface intersected by any plane, (1) perpendicular to a common axis (cylinder, cone), or (2) passing through a common center (sphere), are equidistant from the axis. A roundness tolerance specifies a tolerance zone bounded by two concentric circles in that plane within which the periphery must lie. Figs. 164, 165, and 166 show the method of specifying the permissible variation.

5.5.5 Cylindricity Tolerance. Cylindricity is a condition of a surface of revolution in which all elements form a cylinder. A cylindricity tolerance specifies a tolerance zone confined to the annular space between two concentric cylinders within which the surface must lie. Fig. 167 shows the method of specifying the permissible variation. Note that the cylindricity tolerance controls roundness and straightness, as well as parallelism of the elements.

5.5.6 Profile Tolerancing. The elements of profiles are straight lines and curved lines, the latter being either arcs or irregular curves. If the dimensions of these elements, as well as the dimensions required for their location, are individually toleranced, they must be individually inspected. Basic dimensions (without tolerances) are applied to the elements in pro-

Chap. 6 Dimensioning and Tolerancing for Engineering Drawings

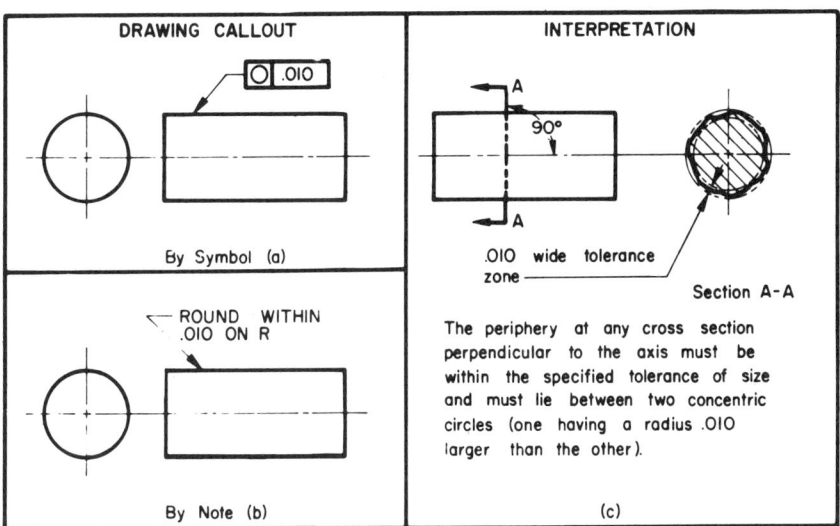

FIGURE 164 – SPECIFYING ROUNDNESS FOR A CYLINDER

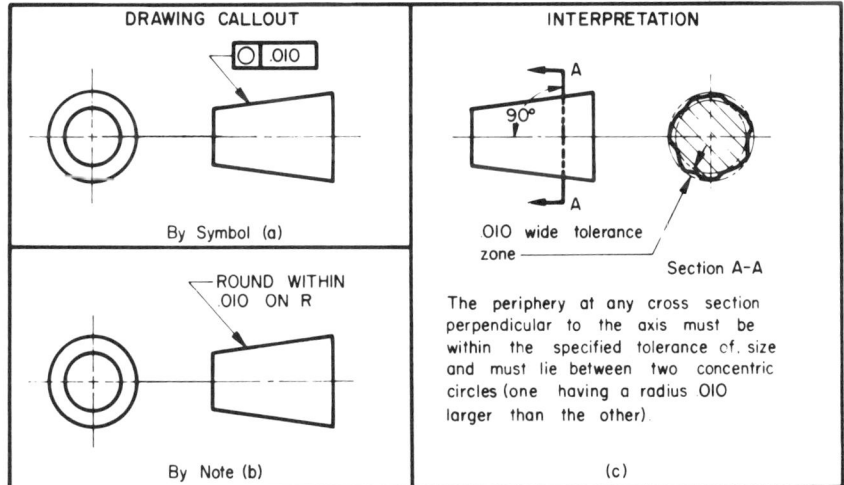

FIGURE 165 – SPECIFYING ROUNDNESS FOR A CONE

110　　　*Dimensioning and Tolerancing for Engineering Drawings*　　Chap. 6

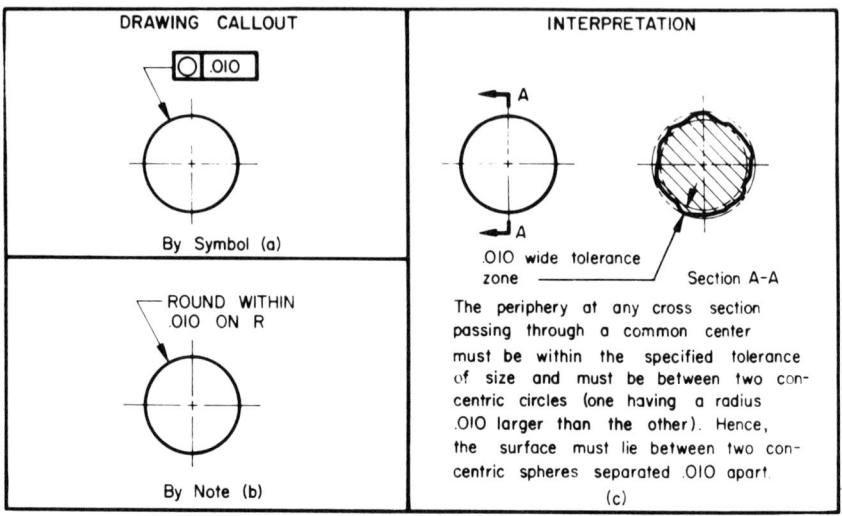

FIGURE 166 – SPECIFYING ROUNDNESS FOR A SPHERE

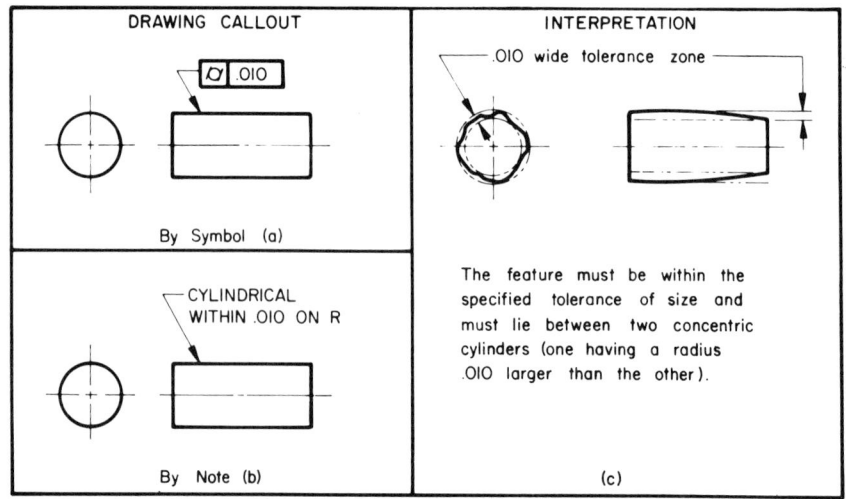

FIGURE 167 – SPECIFYING CYLINDRICITY

Chap. 6 Dimensioning and Tolerancing for Engineering Drawings 111

file tolerancing. This method is used where a uniform amount of variation either unilaterally or bilaterally along a profile and normal to it may be permitted.

5.5.6.1 Application. Where profile tolerances are specified, two parallel boundaries of perfect form are established along the true profile within which the surface or line must lie. For both surfaces and lines, it is necessary to express tolerances in association with the desired profile in a plane of projection on the drawing in the following manner:

a) An appropriate view or section is drawn which includes the desired basic profile.

b) The profile is dimensioned by untoleranced dimensions identified as BASIC. This dimensioning may be in the form of located radii and angles, or it may consist of coordinate dimensioning to points on the profile. Information must be included specifying that general tolerance notes do not apply to the basic dimensions establishing the profile.

c) At a conspicuous location along the profile, a tolerance zone is shown by one or two phantom lines drawn parallel to the profile. The distance between the phantom lines and the basic profile may be greater than the scale value, in order to obtain drawing clarity. The zone may be shown divided bilaterally to both sides of the true profile or unilaterally to either side of the true profile. See Fig. 168.

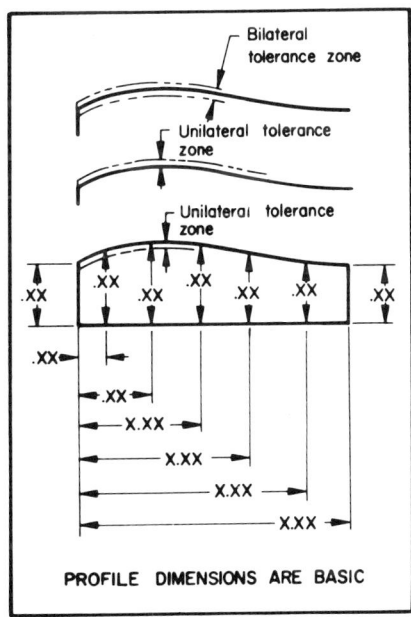

FIGURE 168 – PROFILE TOLERANCE ZONES

d) Appropriate dimensioning is added as well as the applicable feature control symbol or note, specifying profile of a surface, as shown in Figs. 169 and 170, or profile of a line, as shown in Fig. 171. The symbol or note should be applied in a view where the surface or line to be controlled is represented as a profile.

e) Profile tolerances are always applied normal to the profile at all points of the profile.

f) The actual surface or line must lie within the specified tolerance zone and all deviations from true profile must blend in conformance to good workmanship standards.

5.5.6.2 Combined Surface and Line Control. Surface and line controls may be applied to the same feature, where applicable. This would occur where the line elements in one direction need to be controlled more closely than the surface as a whole.

5.5.6.3 Indication of Profile Extremities. If one or more lines or surfaces of a part are profile toleranced and other features are controlled by a general tolerance or individually toleranced dimensions, the extremities of each profile must be clearly indicated. See Fig. 170.

. . .

5.6.5 Runout Tolerance. A runout tolerance establishes a means of controlling the functional relationship of two or more features of a part within the allowable errors of concentricity, perpendicularity and alignment of the

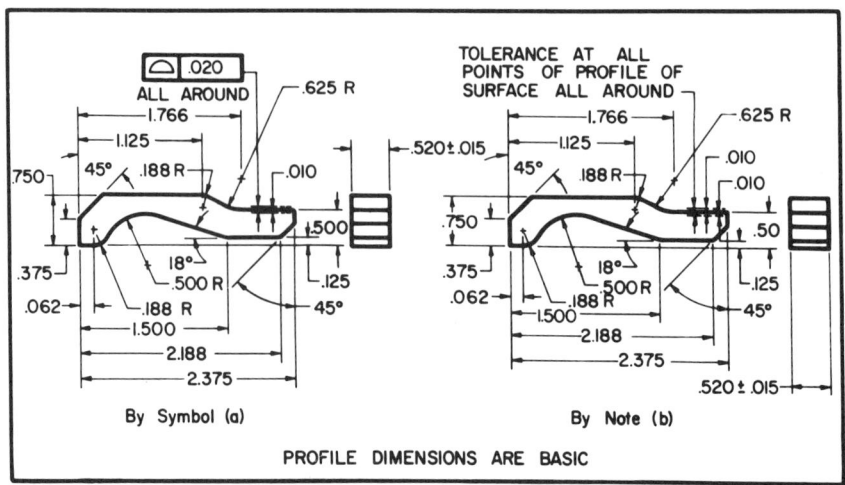

FIGURE 169 – SPECIFYING PROFILE OF A SURFACE

Chap. 6 Dimensioning and Tolerancing for Engineering Drawings

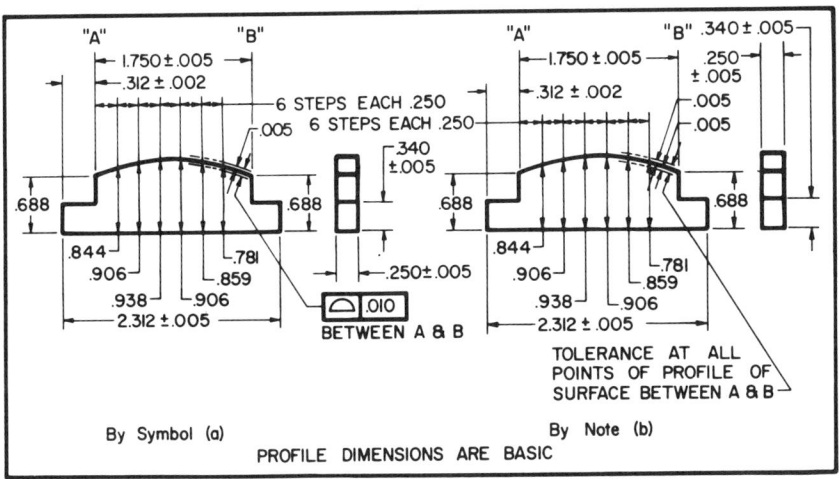

FIGURE 170 – SPECIFYING PROFILE OF A SURFACE

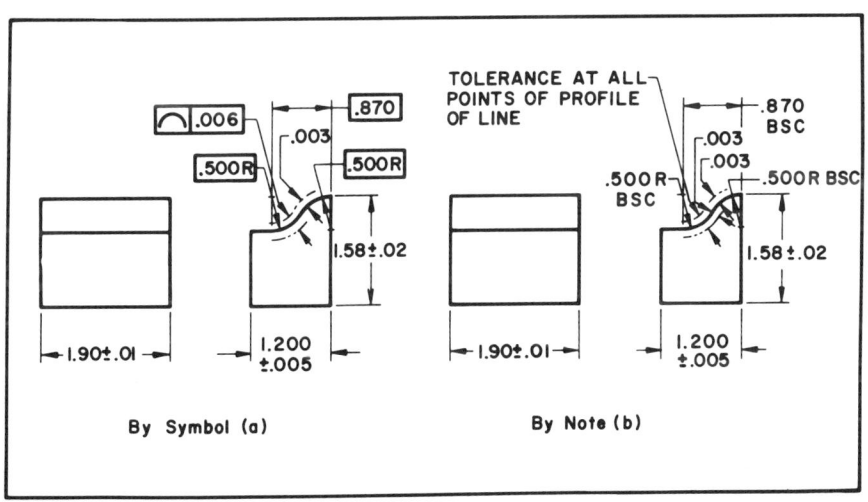

FIGURE 171 – SPECIFYING PROFILE OF A LINE

features. It also takes into account variations in roundness, straightness, flatness and parallelism of individual surfaces. In essence, it establishes composite form control of those features of a part having a common axis. Necessarily then, measurements should be taken under a single setup, and normal to the true, desired geometrical shape. In applying a runout tolerance, the accessibility of datums for single, set-up measurements should be given consideration.

5.6.5.1 Basis of Control. In order to control the relationship of features, it is necessary to establish a datum axis about which the features are to be related. This axis may be established by a diameter of considerable length, two diameters having considerable axial separation or a diameter and a face which is at right angles to it. Insofar as possible, surfaces used as datums for establishing axes should be functional.

5.6.5.1.1 In extreme cases the pitch diameter of a thread, gear or spline, or some such feature, may be desirable for establishing an axis but the use of such features as datums should be avoided. A nearby surface, even though nonfunctional, may be used to better advantage.

5.6.5.2 Basic Control Notes. Where the functional requirements of surfaces are in reference to a common axis, one of the basic form control notes stated below can be specified. The form control is assigned a suitable designating letter and the letter is applied to each related surface. The notes or symbols control concentricity, parallelism and perpendicularity of specified surfaces to the mounting surface or surfaces, and roundness, flatness, parallelism and straightness of each specified surface. Where the basic form control is specified by a feature control symbol, the runout symbol ✓ is used to designate the required form and notes are not used.

5.7 Free-State Variation.
Free-state variation is the amount a part distorts after removal of external forces applied during manufacturing. For instance, parts consisting essentially of shells with a thin wall thickness in proportion to the diameter. Geometric tolerances (such as roundness and concentricity) cannot be properly applied without controlling free-state variation.

5.7.1 Nonrigid Parts. Parts distort if resting in free state on another surface. It is only when such distortion causes parts to fall outside of drawing limits that it must be taken into account on drawings. For purposes of this standard, parts that are subject to such distortion shall be referred to as "nonrigid parts," and such distortion shall be referred to as "free-state variation."

5.7.2 Allowable Free-State Variation. Variations in the free state can exist in two ways:

a) Distortion due to the weight or flexibility of the part.

b) Distortion due to internal stresses set up in fabrication.

These variations to be allowable must be within an elastic range that will allow the part to be brought within drawing tolerance by forces equivalent to those that can be exerted by employing the expected method of assembly. The word RESTRAINED in a drawing note stating an allowance for free-state variation shall be interpreted to mean restraint within the limitations of the preceding sentence when the note does not specify the nature of the restraint.

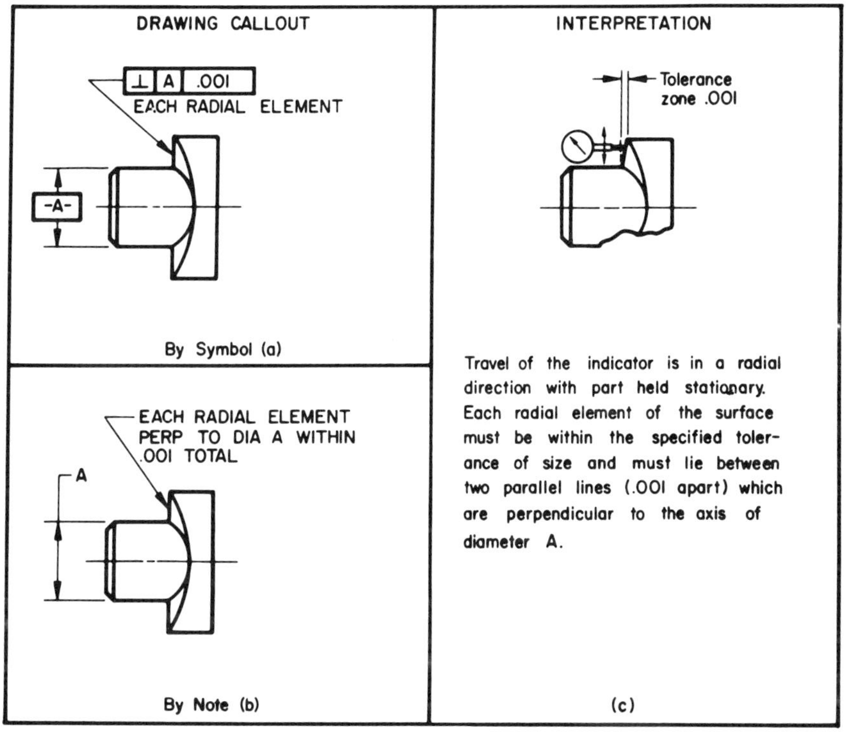

Chap. 6 *Dimensioning and Tolerancing for Engineering Drawings* 115

. . .

DRAWING CALLOUT	INTERPRETATION
⊥ A .001 — EACH RADIAL ELEMENT —A— By Symbol (a)	Tolerance zone .001
EACH RADIAL ELEMENT PERP TO DIA A WITHIN .001 TOTAL A By Note (b)	Travel of the indicator is in a radial direction with part held stationary. Each radial element of the surface must be within the specified tolerance of size and must lie between two parallel lines (.001 apart) which are perpendicular to the axis of diameter A. (c)

FIGURE 189 – SPECIFYING PERPENDICULARITY (RADIAL PERPENDICULARITY)

. . .

5.7.3 Control of Free-State Variation. If part is nonrigid as defined in Paragraph 5.7.1, free-state variation shall be controlled as follows:

a) Select and identify the features (rabbet diameter, bosses, flanges, etc.) to be used as datum surfaces. Since these surfaces may vary from drawing tolerance when they are in free state, it is necessary to specify either their maximum allowable free-state variation, or the maximum force necessary to restrain each of them to drawing tolerance. See Fig. 202.

b) Determine the amount of the restraining and holding forces necessary to simulate expected assembly conditions. Specify on the drawing that if restrained to this condition, the remainder of the part, or certain features thereof, shall be within stated tolerances. See Fig. 202.

It should be noted that if the dimensions and tolerance are met in the free state, it is not necessary to restrain the part.

DIA A SHALL BE ROUND WITHIN .10 ON R IN FREE STATE. WHEN DIA A IS RESTRAINED TO DRAWING TOLERANCE AND SURFACE B IS HELD FLAT WITH THE EQUIVALENT OF 80 TO 130 LB-IN TORQUE AT EACH OF 64 BOLT HOLES, USING .250-28 UNF-3A BOLTS, WITH PART IN HORIZ POSITION; DIA C SHALL BE CONCENTRIC TO DIA A WITHIN .06. DIA C SHALL BE ROUND WITHIN .04 ON R AND ALL BOSSES SHALL BE WITHIN DRAWING TOLERANCE.

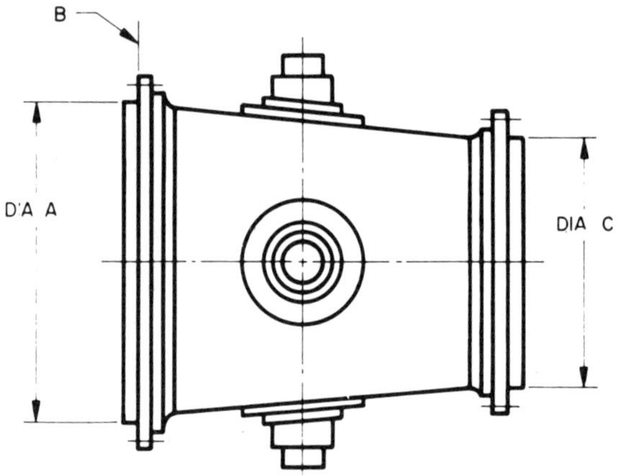

FIGURE 202 – SPECIFYING FREE-STATE VARIATION

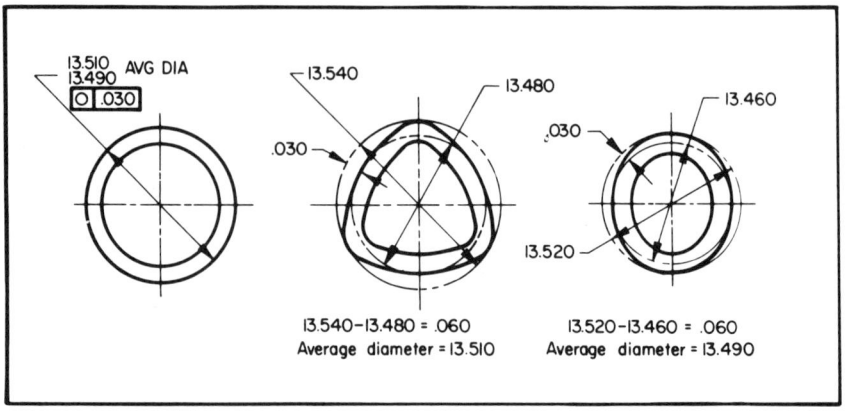

FIGURE 203 – SPECIFYING FREE-STATE VARIATION

5.7.4 Average Diameter. An average diameter is the mean of several diameters (not less than four) across a circular or spherical part used to determine conformance to diameter tolerance only. If practicable, the average diameter may be determined by using a periphery tape. Only when a diameter is allowed a maximum roundness tolerance in the free state should it be labeled AVG DIA. Free-state variation for simple parts may be controlled by specifying the roundness tolerance (by symbol or note) and by labeling the diameter AVG DIA. See Fig. 203.

APPENDIX A

DATUMS—THE THREE-PLANE CONCEPT

A1 General. The purpose of this appendix is to promote a common understanding and interpretation of datum dimensioning by explaining how parts are related to a reference system composed of three mutually perpendicular planes. This explanation is intended to supplement the references made to datums in other sections of the standard.

A2 Terms. The term datum has long been used in a general sense to indicate one end of a relationship between a toleranced feature and other features of a part. However, to permit a valuable distinction between an actual feature and its geometric counterpart, a more precise terminology is required. Part features are therefore called datum features, while the geometric counterparts with which they are associated are called datum points, lines, planes, cylinders, axes, etc.

A3 Three-Plane Reference System. The designer selects as origins for dimensions those surfaces or other features most important in the functioning of a part. Enough features are chosen to position the part in relation to a set of three mutually perpendicular planes, and all related measurements on the part are then made from these planes. See Fig. A1.

    A3.1 Datum Planes. Obviously, however, measurements cannot be made from theoretical planes. The planes are therefore assumed to exist, not in the part itself, but in the much more precisely made manufacturing

Chap. 6 Dimensioning and Tolerancing for Engineering Drawings   119

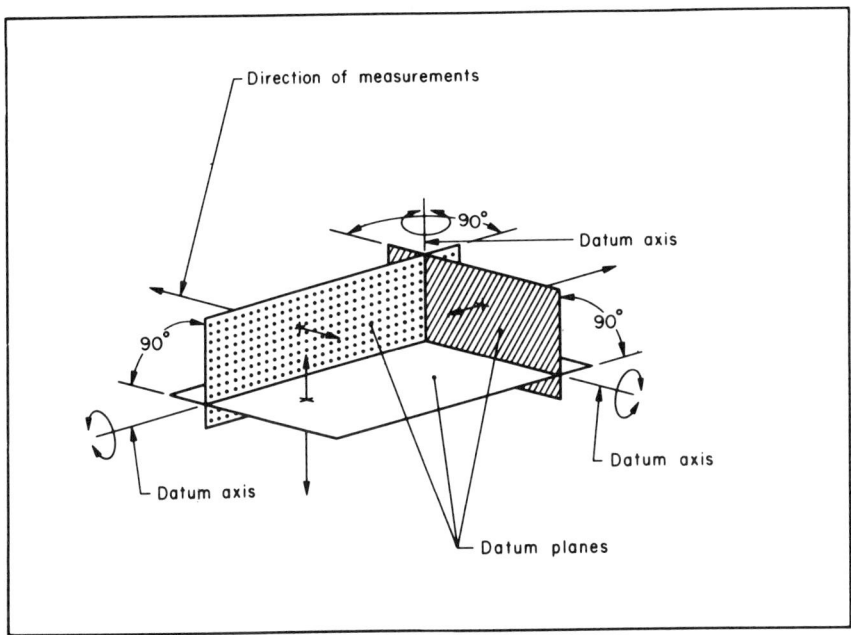

FIGURE A1 – THREE-PLANE REFERENCE SYSTEM

or inspection equipment. Machine tables, surface plates, etc. are not true planes, but they are usually of such high quality that they simulate true planes adequately. It is worth noting that the selection of three mutually perpendicular planes as a reference system is not arbitrary, for most manufacturing and precision measuring equipment operate on the standard coordinate system. Measurements, then, are made from planes or axes in the processing equipment. See Figs. A2 and A3.

A3.2 Association With Datum Features. When magnified, flat surfaces of manufactured parts are seen to have geometric irregularities. They therefore make contact with a datum plane at a number of points, as shown in Fig. A4. This means that in order to apply the three-plane concept it is necessary to determine the minimum number of high points that will establish the set of three mutually perpendicular planes, or in other words, the number of such points that would have to touch the three planes if placed in contact with them.

A3.2.1 By geometry, the first plane is determined by three points not in a line. With this plane determined, a second plane perpendicular to it is determined by two points, and a third plane perpendicular to the first and second is determined by one point. That is to say, a minimum of six high points on a rectangular part determines the location of the three planes.

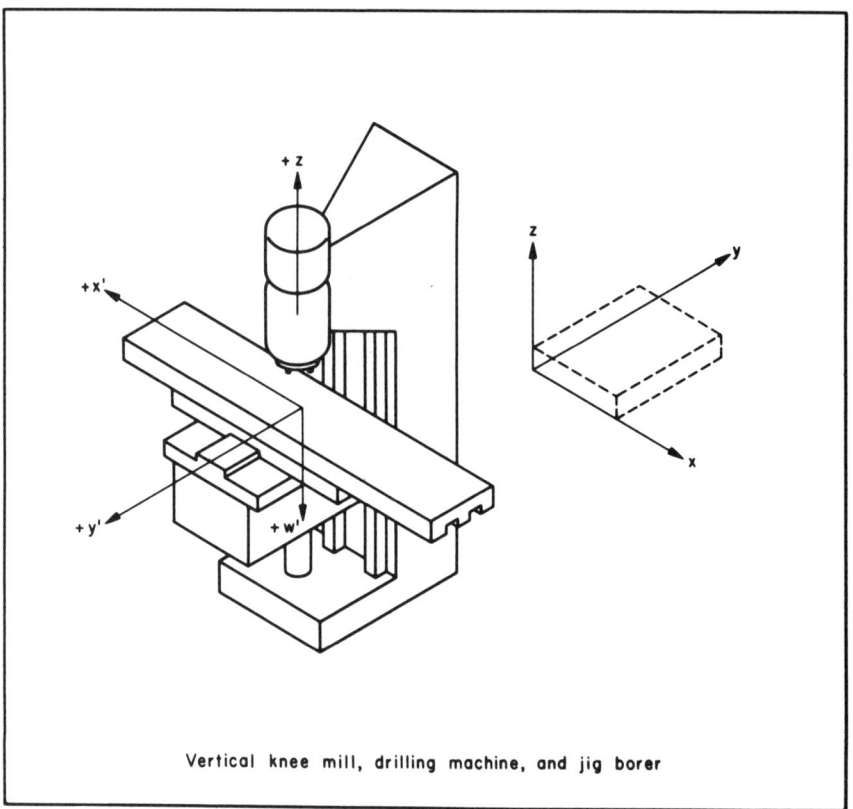

FIGURE A2 - MACHINE TOOL AXES AND MOTIONS APPLICABLE TO MANUFACTURE OF RECTANGULAR PARTS

A4 Sequence of Datum Features. These datum planes have been referred to as first, second, and third, but of course no such sequence exists among the planes in the processing equipment. Hence, to establish uniformity of interpretation in manufacture and inspection, it is necessary to specify on drawings the desired sequence of datum features. See. Fig. A5.

A4.1 Selection of Sequence. The proper sequence should be selected primarily on the basis of the functional requirements of the design, although manufacturing and inspection methods must also be taken into account. In Fig. A6(a), the datum features are identified as surfaces P, S, and T. As can be seen from Fig. A6(b), all of these surfaces are important for the proper assembly and functioning of the part.

A4.1.1 P, S, and T are the primary, secondary, and tertiary datum features, respectively, and would appear in that order in the drawing callout (by symbol or by note). In this example, three datum features are used.

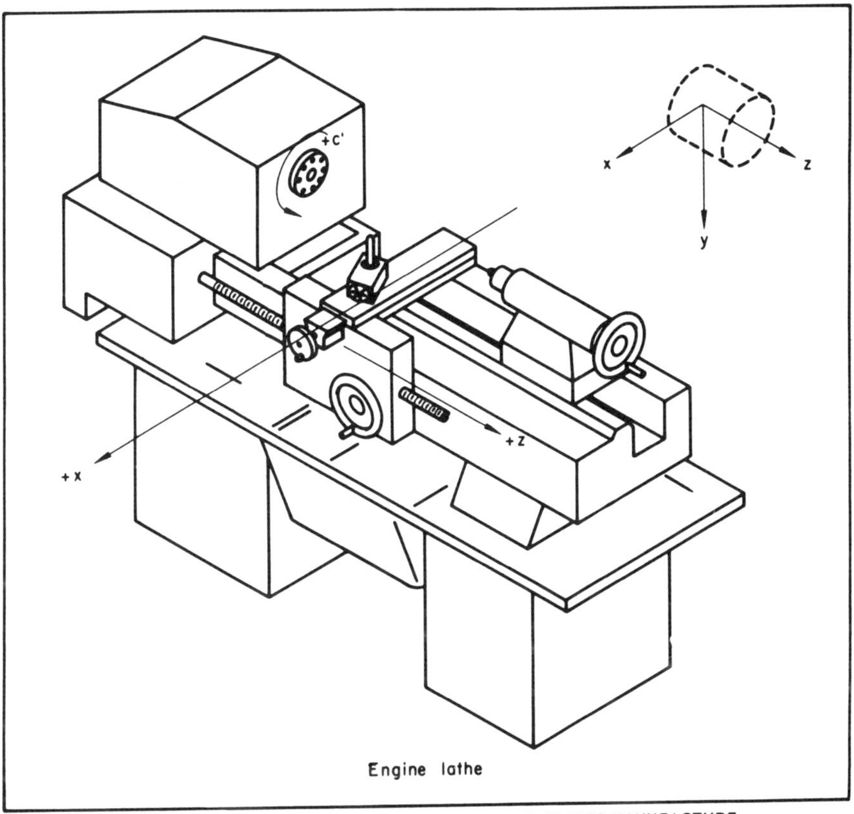

Chap. 6   Dimensioning and Tolerancing for Engineering Drawings   121

FIGURE A3 – MACHINE TOOL AXES AND MOTIONS APPLICABLE TO MANUFACTURE OF CYLINDRICAL PARTS

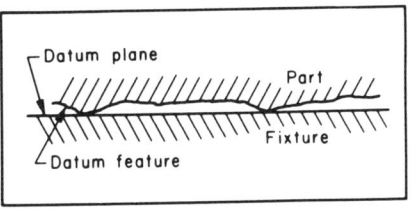

FIGURE A4 – MAGNIFIED SECTION OF FLAT SURFACE IN CONTACT WITH SIMULATED DATUM PLANE

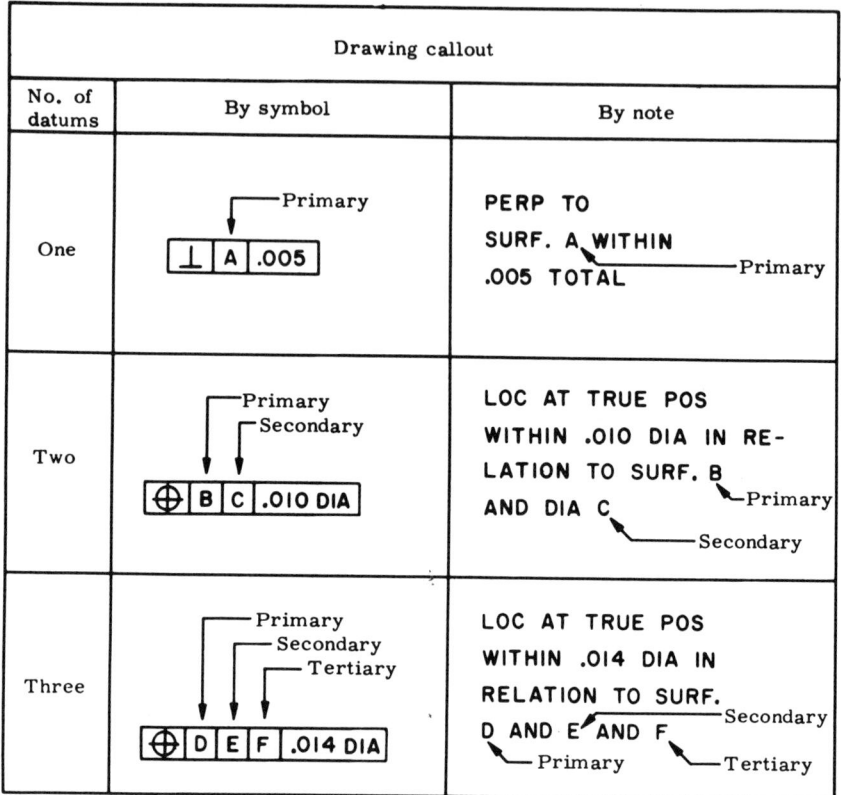

FIGURE A5

However, depending upon the type of tolerance (position or form) and the particular relationship required, it is sometimes necessary to specify only one or two datum features.

A4.2 Sequence Interpretation. The primary datum feature relates the part initially to the three-plane reference framework by bringing a minimum of three high points into contact with the first datum plane, as shown in Fig. A7(a). The part is further related to the framework by bringing at least two points of the secondary datum feature into contact with a second datum plane in Fig. A7(b), and the relationship is completed by bringing at least one point of the tertiary datum feature into contact with the third datum plane. See Fig. A7(c). Since measurements are made from the datum planes, the proper positioning of the datum features in relation to the planes during manufacturing and inspection ensures the desired common basis for measurements.

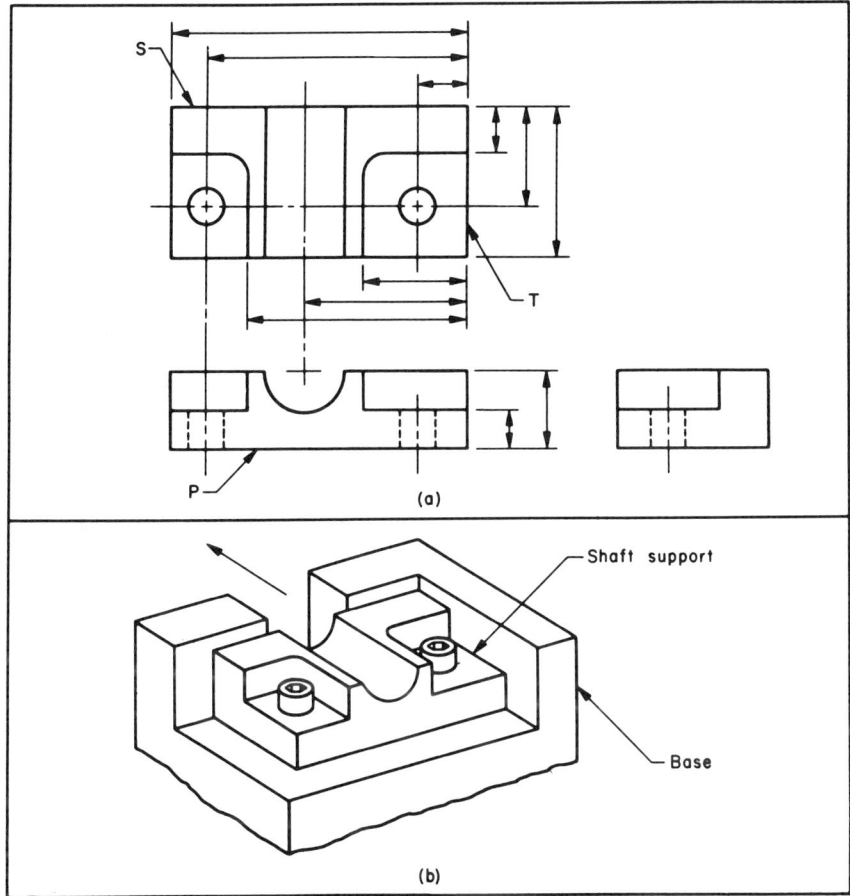

FIGURE A6 – EXAMPLE OF PART WHERE DATUM FEATURES ARE FLAT SURFACES

A5 Parts With Cylindrical Datum Features. As mentioned earlier, the datum planes exist in the manufacturing and inspection equipment. They may be either finished surfaces approximating true planes or planes developed by coordinate movements of the equipment. For a rectangular part, the three planes are easily visualized, since the flat surfaces of the part resemble datum planes. Cylindrical surfaces of parts differ since they bear no resemblance to datum planes. It is possible, however, to visualize two datum planes, usually represented by centerlines at right angles on the drawing, intersecting at the axis of a cylindrical datum feature. Thus, one cylindrical datum feature is associated with two datum planes.

A5.1 Three-Plane Relationship. Application of the three-plane concept to a cylindrical part is best shown by an example such as that of Fig.

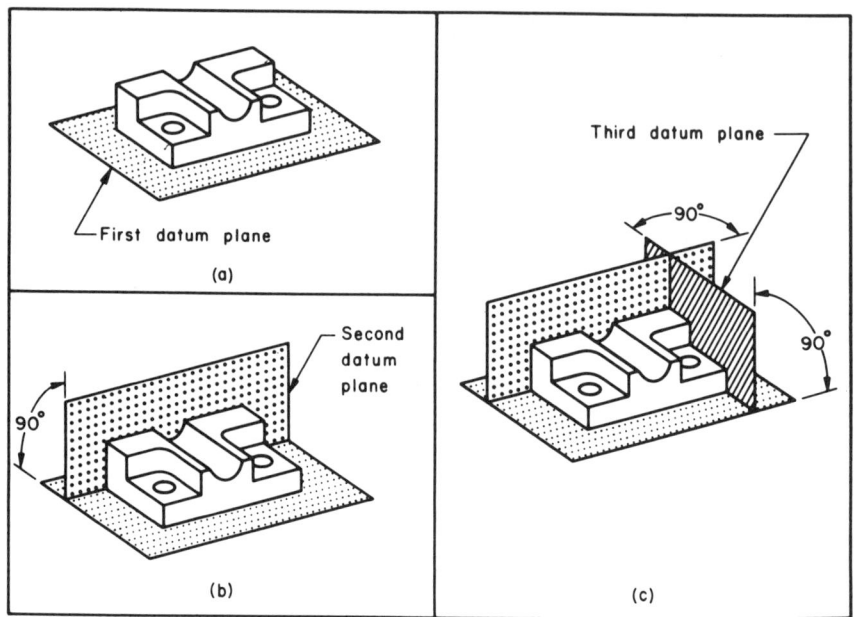

FIGURE A7 – SEQUENCE OF DATUM FEATURES RELATES PART TO THREE-PLANE REFERENCE FRAMEWORK

A8(a). Datum feature K is a sealing surface to which the bolt clearance holes are normal; it is therefore selected as the primary datum feature. The holes are related to a cylindrical datum feature M and are dimensioned from centerlines through the center of M. Since the part has depth, the centerlines represent planes through the part center as shown in Fig. A8(b), and these planes represent the second and third datum planes (L and T). In locating the clearance holes, all measurements are made relative to these three datum planes.

A5.2 Association With Datum Planes. In Fig. A8(c), the machine table represents the first datum plane, and the primary datum feature, surface K, is in direct contact with it. L and T, the second and third datum planes, are developed by longitudinal and transverse movements of the table; they are not represented by physical surfaces and do not come into contact with datum feature M.

A5.2.1 Where a diameter is designated as a datum feature, the line formed by the intersection of two datum planes is called the datum axis. Note that in the example the sequence of planes L and T is immaterial, since rotation of the pattern about the datum axis would not affect the functioning of the part. Consequently, in the drawing callout only two datum features would be specified: (1) primary datum feature K, a flat surface associated with the first datum plane and (2) secondary datum fea-

Chap. 6 Dimensioning and Tolerancing for Engineering Drawings

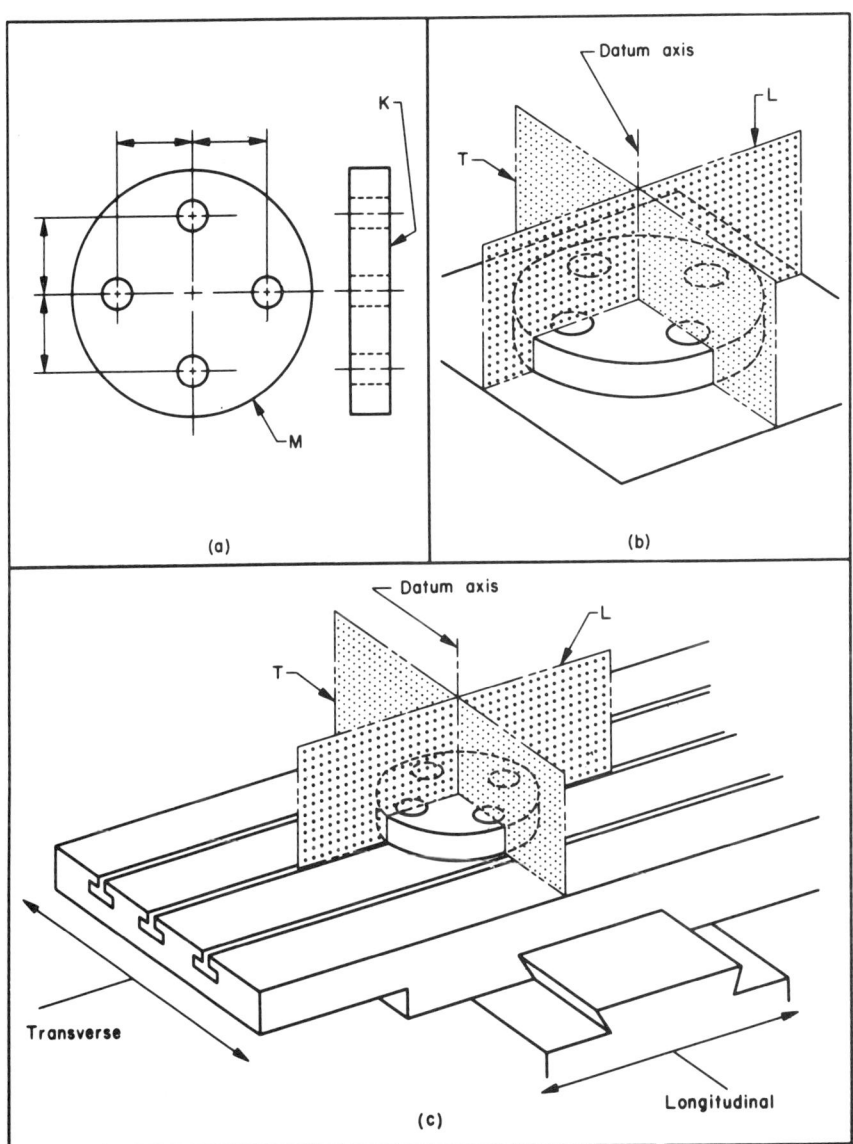

FIGURE A8 – EXAMPLE OF PART WITH CYLINDRICAL DATUM FEATURE

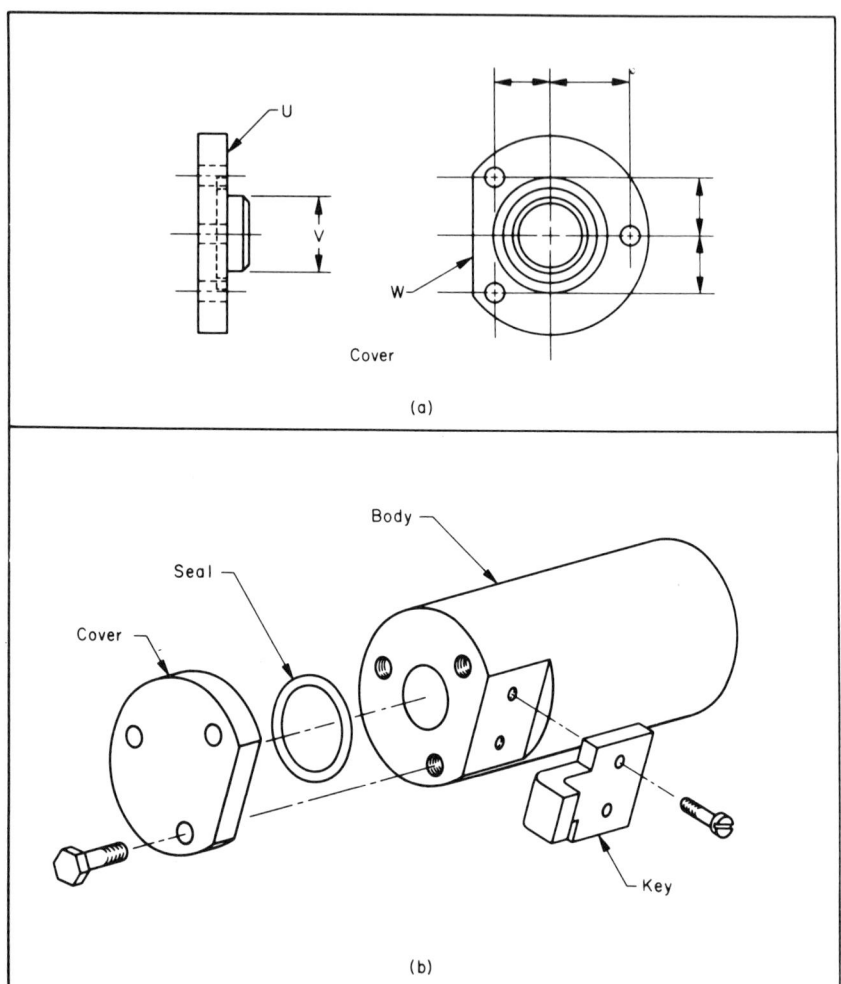

FIGURE A9 - EXAMPLE OF PART WHERE ANGULAR ORIENTATION IS IMPORTANT

ture M, a cylindrical feature associated with the second and third datum planes or datum axis.

A5.3 Angular Orientation. Where it is important to establish the angular orientation of the two planes about the datum axis, a third (or tertiary) datum feature is specified in the callout. This situation is illustrated by the cover in Fig. A9. Surface U is in contact with its mating surface and the clearance holes are normal to it; therefore, it is the primary datum feature. The secondary datum feature is diameter V of the cover, which provides axial alignment with the body. Since V is a diameter, it is associated with two datum planes. However, in this case the flats of both cover and body must align to provide for mounting of the key. Consequently, the angular orientation of the datum planes through V must be established in relation to an auxiliary plane associated with datum feature W and perpendicular to the first datum plane. That is, surface W (the tertiary datum feature in the drawing callout), establishes the auxiliary datum plane required for proper orientation.

A6 Datum Features Subject to Size Variations. It should be noted that datum features such as diameters and widths differ from flat surfaces in two ways. Consequently, the making of measurements from these datum features is somewhat different, although the principle is the same as with flat surfaces. First, dimensions are shown on drawings as originating from centerlines or center planes rather than from the surface of the feature. For example, diameter A of the part shown in Fig. A10(a) is a datum feature, and the measurements indicated by the dimensions shown originate from datum planes which intersect at a datum axis. Second, diameters and widths are subject to variations in size as well as to surface irregularities. However, because a cylindrical datum feature, such as diameter A in Fig. A10(a), varies in size, the application of RFS or MMC must be considered in the definition of such a part. Just as a flat surface is associated with its geometric counterpart (a true plane), a diameter is likewise associated with its geometric counterpart, a true cylinder. The datum axis (the intersection of two datum planes) is the axis of the true cylinder associated with a cylindrical datum feature.

A6.1 Primary Datum Feature RFS. The sequence of the datum reference in a drawing callout has a significant bearing on manufacturing and inspection processes, as the following figures show. Fig. A10(b) illustrates the case in which diameter A is the primary datum feature, RFS is applied, and surface B is the secondary. Since A would be specified first in the drawing callout, the datum axis is the axis of the smallest true cylinder which will contact the high points of surface A. In theory measurements are made from the axis of the true cylinder which contacts the datum feature, but in practice the true cylinder is adequately simulated by processing equipment (chuck, mandrel, centering device, etc.) which simply centers the feature. The axis of the processing equipment therefore becomes the datum axis and is used as the origin for measurements. The first and second datum planes intersect at the datum axis and at least one point of surface B contacts the third datum plane to complete the three-plane relationship.

A6.2 Secondary Datum Feature RFS. The situation is quite different if, as in Fig. A10(c), B is the primary datum feature, A is the secondary, and RFS is again applied. Here the datum axis is the axis of the smallest true cylinder perpendicular to the first datum plane (the plane which contacts

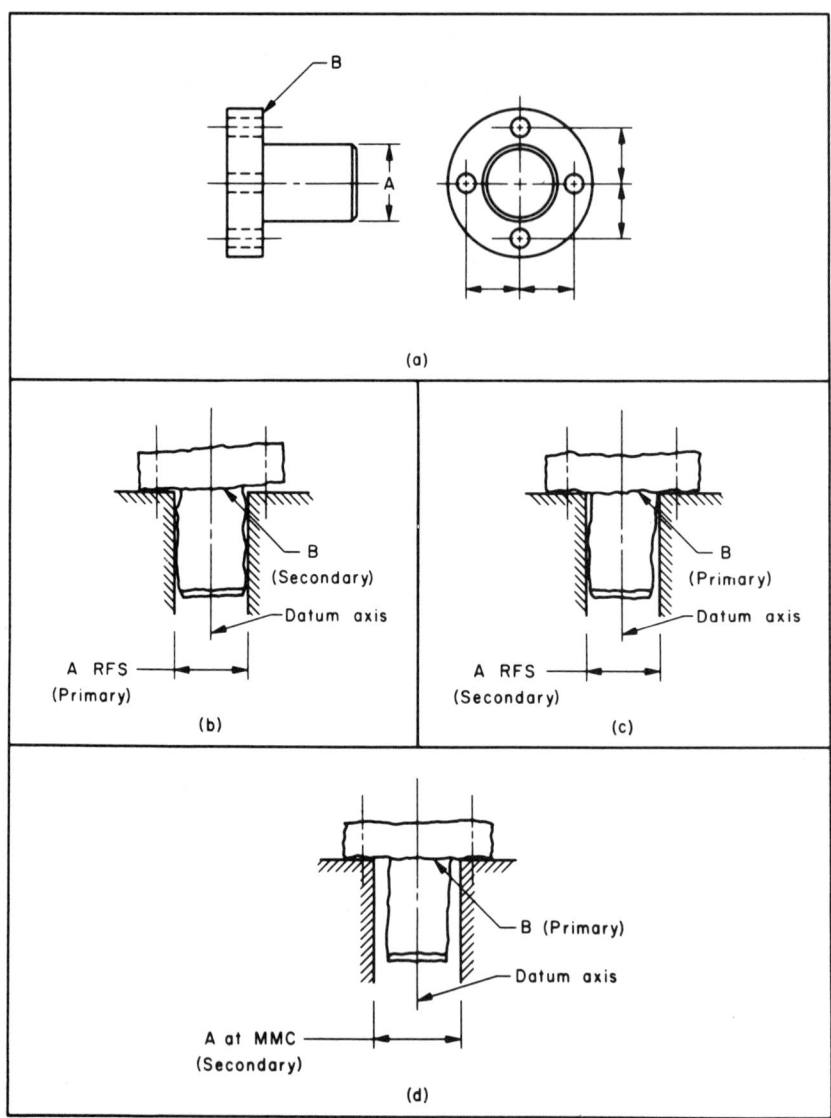

FIGURE A10 – EFFECT OF RFS OR MMC APPLIED TO DATUM FEATURES

at least three high points of surface B) that will contact points of surface A. The second and third datum planes intersect at the datum axis, thereby completing the three-plane relationship.

A6.3 Secondary Datum Feature at MMC. Fig. A10(d) shows the case in which B is the primary datum feature, A is again the secondary, and MMC is applied to datum feature A. Here the datum axis is the axis of a

Chap. 6 *Dimensioning and Tolerancing for Engineering Drawings*   129

true cylinder of MMC size perpendicular to the first datum plane (the plane which contacts at least three high points of surface B). This cylinder is simulated in the processing equipment by a diameter of fixed (MMC) size. Note that where datum feature A is finished away from its MMC, its surface does not necessarily contact the cylinder. The second and third datum planes intersect at the datum axis to again complete the three-plane relationship.

A7 Specifying a Datum Feature in a Pattern of Similar Features. The three-plane concept offers further advantages in certain true position dimensioning applications because, where necessary, it can provide a more precise drawing definition.

A7.1 Manufacturing and Inspection Considerations. Mass production usually involves manufacturing and inspection operations performed in a simultaneous manner, and processing equipment designed for this purpose is used. For example, patterns of holes are often made by using multiple spindle machines or by locating from fixed bushings in drill jigs, and such patterns are inspected for location by using functional type (i.e., fixed-pin) receiver gages. However, there are important exceptions to this procedure. In some cases production quantities are limited; in others, production-destined items are still in the early development phase. That is, parts must be defined, produced, and inspected using existing facilities and equipment. In such cases, parts may be fabricated without the benefit of hard tooling and inspected on an open set-up basis without the benefit of inspection gages.

A7.2 Identifying a Datum Feature. Positional tolerancing a pattern of similar features without specific datum references, although suitable for simultaneous operations and measurements in production work, becomes difficult to interpret in terms of individual dimensions and measurements. Time consuming methods requiring repositioning of the part, repeated measurements, and mathematical calculations are often necessary to verify whether the feature pattern location and the location of features within the pattern are within the limits specified. However, identifying a datum feature in the pattern, as explained in Paragraph 4.6.2, provides a way of relating a pattern of features to a set of three mutually perpendicular planes (the standard coordinate system). Once this relationship has been established, individual measurements can conveniently be made.

A7.3 Compatibility With End Product Requirements. Note that this method is compatible with the simultaneous operations encountered in production; it requires no further qualification of the drawing for this later phase. The "end product" drawing concept, upon which this standard is based, allows the manufacturer to simultaneously drill all holes in a pattern if he wishes, provided the end requirements specified have been met.

A7.4 Referee Measurements. In addition to clarifying gaging concepts for the gage designer, datum references provide a sound basis for the making of "referee" measurements where the conformance of "marginal" parts to drawing requirements must be verified.

A8 Datum Targets

A8.1 Qualification of Datum Surfaces. The datum features of certain parts frequently require further qualification before they can be related to the three-plane framework. Examples are surfaces produced by casting, forging, and molding; surfaces adjacent to welding; and thin sheet metal. All of these are subject to bowing, warping, and distortion. Since they are not reasonably flat, it is impractical to designate an entire surface to be associated with a datum plane; accurate and repeatable measurements simply cannot be made on such surfaces. For instance, a cast surface placed in contact with a surface simulating a datum plane (a machine table or surface plate, for example) may actually rock.

A8.2 Datum Target Method. The datum target method is a useful technique for relating such parts to the three-plane system. Points or areas are selected and designated by datum target symbols on drawings. It is at these places that contact is made with the processing equipment. The geometrical considerations discussed earlier are required to determine the number of points or areas associated with each plane. After this has been done, all measurements can be made from common origins. Datum targets are usually located by untoleranced dimensions, and the accuracy of tool or gage feature locations representing datum targets is held within standard tool and gagemaker tolerances or controlled by general workmanship requirements.

APPENDIX C

FORMULAS FOR TRUE POSITION TOLERANCING OF ROUND HOLES

C1 General. The purpose of this appendix is to give formulas for use in assigning positional tolerances to round holes. These formulas are based on equal positional tolerances for adjoining parts. (Unequal tolerances are discussed in paragraph C4.)

C2 Formulas. The following formulas will give a "no-interference, no-clearance" fit at maximum material limits of clearance holes:
a) For diameter method:
Where T = permissible true position tolerance based on diameter method
H = minimum diameter of clearance hole
F = maximum diameter of fastener

1) Retaining Hole and Clearance Hole. Where one part has a hole (threaded or plain) that retains a part such as a bolt, stud, or tight-fitting dowel, and the adjoining part has a clearance hole (for the bolt, stud, or dowel), the following formulas may be used:

$$T = \frac{H - F}{2}, H = F + 2T$$

2) Bolt Clearance Holes. Where adjoining parts have the same basic clearance between holes and fasteners, the following formulas apply:

$$Y = H - F, H = F + Y$$

b) For radius method:

where Z = permissible true position tolerance based on radius method
H = minimum diameter of clearance hole
F = maximum diameter of fastener

1) Retaining Hole and Clearance Hole. Where one part has a hole (threaded or plain) that retains a part such as a bolt, stud, or tight-fitting dowel, and the adjoining part has a clearance hole (for the bolt, stud, or dowel), the following formulas may be used:

$$Z = \frac{H - F}{4}, H = F + 4Z$$

2) Bolt Clearance Holes. Where adjoining parts have the same basic clearance between holes and fasteners, the following formulas apply:

$$Z = \frac{H - F}{2}, H = F + 2Z$$

C3 Out-of-Squareness of Threaded Holes. Where the above formulas are used to derive tolerances for threaded holes, or holes for tight-fitting members such as dowels, the X and Y values should be specified in conjunction with a "projected tolerance zone" value, given on the drawing; otherwise, fastener interference may occur. See Paragraph 4.7.

C4 Unequal Tolerances. Although the above formulas are based on equal positional tolerances for adjoining members, equal tolerances are not always desirable. It may be preferable in some cases to divide the available tolerance unequally; for instance, it is sometimes more practicable to assign a larger positional tolerance to threaded holes in one part, and a smaller tolerance to the corresponding clearance holes in the adjoining part.

BIBLIOGRAPHY

Books
Concepts of the True Position Dimensioning System (Chicago, Ill.: Central Scientific Company).
Fundamentals of Tool Design, 1st ed. (Englewood Cliffs, N.J.: Prentice-Hall, 1962).
Gladman, C. A. *Manual for Geometric Analysis of Engineering Designs* (Australia: Publicity Press Ltd., 1966).
Liggett, John V. *Fundamentals of Position Tolerance*, 1st ed. Edited by Ivan R. Vernon. (Dearborn, Mich.: Society of Manufacturing Engineers, 1970).
Roth, Edward S. *Functional Inspection Techniques*, 1st ed. Edited by Robert R. Runck. (Dearborn, Mich.: American Society of Tool and Manufacturing Engineers, 1967).

Pamphlets and Periodicals
Black, F. W., Jr., "An Aerospace Industry Report on TPDT," *Machine Design* (March 20, 1969).
Roth, Edward S. "Don't Reject Those Good Parts," *The Tool and Manufacturing Engineer*, Vol. 51, No. 3 (September, 1961).
———. "Functional Gaging with Optical Projectors," *The Tool and Manufacturing Engineer*, Vol. 42, No. 4 (May, 1966).
———. "How to Make Temporary Functional Gages," *The Tool and Manufacturing Engineer*, Vol. 52, No. 3 (March, 1966).
———. "Lets Stop Interpreting," *Quality Assurance*, Vol. 4, No. 1 (January, 1966).

——. "Paper Layout Gaging," *The Tool and Manufacturing Engineer*, Vol. 58, No. 3 (March, 1967).

Industry and Government Reports

Foster, L. *A Treatise on Geometric and Positional Dimensioning and Tolerancing* (Minneapolis-Honeywell, 1963).

"Positional Tolerancing at Sandia Corporation," *Paper SCR-83A* (Albuquerque, N.M.: Sandia Corporation, May, 1959).

Roth, Edward S. "Design and Inspection Factors That Affect Sensible Manufacturing," *Paper SCR-496* (Albuquerque, N.M.: Sandia Corporation).

——. "Gages Reject Too High a Percentage of Good Parts," *Paper SCR-291* (Albuquerque, N.M.: Sandia Corporation).

——. "How Positional Tolerancing Clarifies Design Intent and Reduces Product Cost," *Paper SCR-154* (Albuquerque, N.M.: Sandia Corporation).

——. "Inspection and Gaging of Positionally Toleranced Parts," *Paper SCR-203* (Albuquerque, N.M.: Sandia Corporation).

——. "Special Gage Designs for Positionally Toleranced Parts," *Paper SCR 292* (Albuquerque, N.M.: Sandia Corporation).

——. "The Phantom Gage," *Paper SCR-617* (Albuquerque, N.M.: Sandia Corporation).

Utter, R. F., et al. *Concepts of the True Position Dimensioning System* (Albuquerque, N.M.: Sandia Laboratories, 1963).

Standards

American National Standards Institute, "Dimensioning and Tolerancing for Engineering Drawings," *Std-ANS Y14.5* (1966).

British Standard, "Engineering Drawing Practice," *Std-308* (London: British Standard House, 1953).

Canadian Standards Association, "Mechanical Engineering Drawing Standards," *B78.1* (1954).

U.S. Military Standard, "Dimensioning and Tolerancing," *MIL-Std-8* (U.S. Government Printing Office, 1959).

U.S. Military Standard, "Gage Inspection," *MIL-Std-120* (U.S. Government Printing Office, 1950).

Transactions and Proceedings

Roth, Edward S. "Consider the Inspection Method When Reviewing the Design," *Western Regional ASQC Conference* (Portland, Oregon, 1964).

——. "Design and Inspection Factors that Affect Sensible Manufacturing," *ASTME Conference on Sensible Manufacturing* (University of Colorado, 1962).

——. "Drawing Standards and Interpretations," *CIRP International Conference on Manufacturing Technology* (University of Michigan, 1967).

——. "The Control of Coaxial Part Features," *ASTME National Convention* (May, 1969).

——. "The Gray Area of Intent," *19th Annual Meeting, Standards and Metrology Division* (American Ordnance Association, 1964).

Manuals

"Tooling Points and Datum Planes," *Boeing Design Manual,* Vol. 2, No. 8, Section 9:32 (Boeing Aircraft Company, 1961).

INDEX

A

ANS Y14.5
   selected portions, 83–132
   symbols, 6–8

B

Basic, definition, 2
Basic interchangeability gages, 27–29
British Standards symbols, 6–8
BSC (*see* Basic)
Buttons, toolmakers', 13

C

Canadian Standards symbols, 6–8
Cartesian coordinates, 9–10
"Chain" dimensioning, 9, 17
Chart gages, 71–74
Clearance hole patterns, 30–32
Coaxiality, definition, 2
Concentricity tolerance
   definition, 2
   gaging for, 75–77
Counterbore patterns, 34–37
Critical part datum features, 43–51
   MMC part datum features, 48
   RFS part datum features, 43
Cylindrical parts, gaging of, 60–63
Cylindrical tolerance zone, 17–18

D

Datum allowance, definition, 2
Datum axis, definition, 2
Datum dimensioning, 9
Datum feature, 4, 10–12
Datum features related to primary datum plane, 57–58
Datum planes, definition, 2
Datum reference planes, 9–10, 12
Dowel pin patterns, 39–41

E

Elements, gage, 12–16
External feature patterns, 38–42
   dowel pins, 39–41
   studs, 38–39
   threaded studs, 41

F

Feature control symbols, 6–8
   box, 6, 7
Feature location and relation gaging
   critical datum features, 43–51
   critical datum features with independent hole pattern, 55
   cylindrical part with two pin patterns, 60–61
   datum features related to primary datum plane, 51
   definition, 3
   independent hole patterns, 51, 55
   non-critical part datum features, 68
   radial patterns of pins and slots, 63–66

Index

two datum features with two feature patterns, 63
Feature relation gages, 30–42
   definition, 3
   external feature patterns, 38–42
   internal feature patterns, 30–38
Fixed-element functional gage, definition, 3
Fixed gage-element, 3, 43–48
Fixed-nut retainer patterns, 37–38
Fixture gage, definition, 3
Floating-nut retainer patterns, 38
Form tolerances, 78–82
Functional gaging principles, 5–6

G

Gage centering element, definition, 3
Gage datum elements, 12–16
Gage element, 3
Gage
   chart, 71–74
   feature location, 3
   feature location and relation, 3
   fixed-element functional, 3
   fixture, 3
   go, 3, 20–22
   interchangeability, 27–29
   not-go, 4, 20–22
   optical chart, 71–74
   phantom, 69–74
   plug, 20–22
   principles, 5–6
   rail, 13
   ring, 20–22
   shake, 3
Gaging
   clearance hole, 31–32
   counterbore, 34–37
   critical datum feature, 43–48, 51–57
   cylindrical part with two-pin patterns, 60–63
   datum features related to primary datum plane, 57–58
   dowel pin pattern, 39–41
   feature relation, 30
   fixed-nut retainer pattern, 37–38
   floating-nut retainer pattern, 36
   independent-hole pattern, 51, 55–57
   less critical part datum feature, 48–51
   maximum material condition, 48–51
   radial patterns of pins and slots, 63–66
   regardless of feature size, 43–48
   stud pattern, 38–39
      threaded, 42
   tapped feature pattern, 32–34
   threaded stud pattern, 42
   three-hole pattern, 58–60

two datum features with two feature patterns, 63
Go gages, 3, 20–22

H

Hole patterns, gaging of
   clearance, 30–32
   counterbored, 34–37
   cylindrical parts, 60–63
   independent, 51, 55
   tapped holes, 32–34

I

Independent hole patterns, 51, 55–57
Internal feature patterns, 30–38
   clearance holes, 30–32
   counterbores, 34–37
   fixed-nut retainers, 37–38
   floating-nut retainers, 36
   tapped holes, 32–34

L

Less critical part datum features, 48–51

M

Mating part thickness, 32–33
Maximum material condition, 24 *(table)*, 25–26
   definition, 3
   part datum features, 48
Mil-Std-120, 5, 20
Minimum material condition, 4
MMC *(see* Maximum material condition)

N

Non-critical part datum features, 68
Not-go gage, 4, 20–22
Nut retainers, gaging of, 36–38

O

Optical chart gaging, 71–74

P

Part datum feature, 10–13
   critical, 43–51
   identification, 9
   not-critical, 68
Part thickness, effect on gage, 32–33
Perpendicularity, gaging for, 79–82
Phantom gaging, 69–74
Plug gage, 20–22
Positional tolerance
   definition, 4
   hole size, 26, 27 *(table)*

maximum material condition, 24 *(table)*, 25–26
 principles, 17–19
Primary datum feature, definition, 4
Profile tolerance, definition, 2

R

Radial patterns, 63–66
Rail gage elements, 13
Receiver gaging principles, 5–6
Regardless of feature size, 23 *(table)*, 24
 definition, 4
 part datum feature, 43–48
RFS *(see* Regardless of feature size)
Ring gages, 20–22

S

Secondary datum feature, definition, 4
Squareness tolerance, definition, 5
Standards
 ANS Y14.5, 6–8, 83–132
 British, 6–8
 Canadian, 6–8
 International Organization for Standards, 6–8
 Mil-Std-120, 5, 20
Straightness tolerance
 definition, 4
 gaging for, 78–79

Stud patterns, 38–39, 42
 threaded, 42
Surface plate inspection, 14
Symbols
 American, 6–8
 British, 6–8
 Canadian, 6–8
 International, 6–8
Symmetry tolerance, 5

T

Tangent part locators, 13
Tapped hole patterns, 32–34
Taylor's principle, 21–22
Tertiary datum feature, definition, 4
Threaded stud patterns, 42
Tolerance
 concentricity, 2
 form, 78–82
 positional, 4, 17–19, 24–27
 profile, 2
 squareness, 5
 straightness, 4
 symmetry, 5
 zone, 17–18
Tolerance zone projection, definition, 4
True position
 definition, 5
 principles, 17–19

Other Books from SME . . .

MANUFACTURING DATA SERIES

Adhesives in Modern Manufacturing

Cold Bending and Forming Tube and Other Sections

Cutting and Grinding Fluids: Selection and Application

Cutting Tool Material Selection

Design of Cutting Tools: Use of Metal Cutting Theory

Functional Inspection Techniques

Fundamentals of Position Tolerance

Gundrilling, Trepanning, and
  Deep Hole Machining (Revised Edition)

High-Velocity Forming of Metals (Revised Edition)

Machining the Space-Age Metals

Non-Traditional Machining Processes

Pneumatic Controls for Industrial Application

Premachining Planning and Tool Presetting

Producibility/Machinability of Space-Age and
  Conventional Materials

Realistic Cost Estimating for Manufacturing

Tool Engineering: Organization and Operation

NUMERICAL CONTROL SERIES

Introduction to Numerical Control in Manufacturing

MANUFACTURING MANAGEMENT SERIES

Introduction to Manufacturing Management

Modern Aspects of Manufacturing Management:
  Selected Readings

Organization for Manufacturing

For further information, write to:

Society of Manufacturing Engineers
Publication Sales Department
20501 Ford Road
Dearborn, Michigan 48128

DISCARDED

JUN 2 4 2025